SAT-1 数学轻松突破 800 分：
思路与技巧的飞跃

时　坚　编著
言　中

中国人民大学出版社
·北京·

本书设计和与众不同之处

自从 SAT 考试改革以来,数学部分就一直是大家热议的话题,很多同学都会问这个问题:新 SAT 的数学部分是简单了还是更难了呢?其实,细细体会一下就会发现这是一个伪命题,有着明显的逻辑错误。题目的难易程度与大家的得分并没有直接的必然联系,国内高考数学或者奥数竞赛的题目很难,依然有同学能做对,而新 SAT 数学貌似容易,初次接触或许就取得了 750 分,但是几次刷分下来,却依然还是这个分数,或者略有提高,有时候还会下降。就是这个"貌似容易"的数学部分,让很多同学与 800 分失之交臂,从而影响了总分,甚为可惜。

为什么会有这样的现象呢?这是因为很多同学包括家长对数学有认识上的误区,经常从所谓的朋友、亲戚那里道听途说,声称某中国学生在国内读书时数学成绩马马虎虎,到了国外后摇身一变,成为了班上数学课的"学霸"。我们相信这样的现象的确存在,特别是中国学生强大的计算能力会让他们的美国老师和同学刮目相看(比如咱们烂熟于心的九九乘法表,但是英语中没有这样的口诀对应,根本就是背不下来嘛)。但是值得注意的是国内的数学课与美国的数学课相比,由于受到历史、文化甚至价值观的影响,在教学的组织与思维方式上有着很大的差别,而这个差别在新 SAT 数学上体现得尤为明显。

新 SAT 数学包括了两个部分,第一部分不能使用计算器(共计 20 题,限时 25 分钟),考查学生对基本概念的理解。由于不能使用计算器,所以这一部分的计算难度不是很大。第二部分可以使用计算器(共计 38 题,限时 55 分钟),考查学生借助工具迅速地实现对数据分析和处理的能力。由于可以使用计算器,所以也就没有什么计算是搞不定的。但是,统计结果表明老 SAT(2016 年 3 月前的考试)数学每题平均只有 27 个单词,新 SAT(2016 年 3 月开始)数学每题平均达到了 51 个单词。可见,改革后的数学对大家阅读能力的要求有了显著的提高,读入的是文字描述,输出的则是数学关系,我喜欢把这称为"转化数学"。这说明新 SAT 数学并不仅仅考查大家的计算能力,更是考查数学思维,即如何用数学的视角解决现实中的问题。之所以很多同学拿不到满分,不是你不会计算,也不是对某个知识点的遗忘,而是没有真正明白新 SAT 数学的"心"!

新 SAT 数学满分为 800,占据了 SAT 考试总分的一半(总分为 1,600),如果放弃获取数学部分的满分,大家似乎有些于心不忍,但投入大量时间把市面上所有的新 SAT 数学题做一遍又心有不甘。正是在广大考生面临着这样的两难抉择时,我们突然有了一个想法——能否以我们对新 SAT 数学的理解,对已有 OG 第二版 8 套题和 14 套亚太和北美考过的真题进行深入研究,总结授课时学生(坚果)的反馈,为大家提供一份高效的复习资料呢?这便是这本书的由来。

今天这本书终于要面世了,我们从构思、收集真题开始至今花了接近两年的时间。其中大部分的时间是用在了"顶层设计"上。什么是高效的复习资料呢？我们认为应该体现三个特性——针对性、多面性和过程性。所谓针对性，是指我们的题目针对中国学生的特点和问题，帮助大家更好地理解新SAT数学考试的真谛，特别是从文字描述中找到数学关系（详见本书第一部分）。所谓多面性，是指我们的题目既有易错题、巧解题，也有强化题和模考题，满足不同层次学生的需求。所谓过程性，是指我们的题目包括了知识清单（入门阶段）、必做题与全真模考（强化阶段）及考前复习方案（冲刺阶段），全程助力考生踏实走好每一步。

我们不提倡题海战术，但是"批判的武器还需要武器的批判"，数学思维与做题的技巧不是说出来的，需要一定的题目来修炼内功。我们从来不搞所谓的考前预测与押题，本书以OG最新第二版8套题和14套真题为蓝本，所有题目均经过精挑细选，希望以此为考生揭示解题规律，突出思维过程，培养做题技巧，真正实现实力的提升。

时　坚　　言　中

2017年7月于南京坚果教育

本书的使用方法

1. 知识点清单包含了新 SAT 数学所涉及的考点，基础薄弱的同学可以先从这部分开始。

2. 第一部分主要针对中国学生普遍感觉吃力的文字题而设定，基础较好的同学可以针对性地开展学习与训练。

3. 对于第二部分易错题分析和第三部分简单粗暴解析法，同学们可以和第四部分数学必做 100 题相互配合使用，通过一定强度的练习，巩固知识点，发现自身问题。

4. 第一至三部分完成后可以开始完成全真模考题。第一套模考题在出题思路、考点顺序、难易程度等方面与真考题高度一致，建议完成第一套模考题时不用严格计时，主要熟悉做题流程。第二套题难度略高于真考题，建议完成第二套模考题时严格计时，体会考场感受。

5. 第六部分提供了考前冲刺复习方案，供考生合理高效安排好临考前一周的复习计划，做到事半功倍。

6. 本书附录 1 是知识点清单，便于坚果们考前查漏补缺；附录 2 是常见数学表达词汇，中英文对照，建议通过坚果教育小助手的每日词测，上课前熟背，对于阅读题干、提高解题速度会大有帮助；附录 3 是数学计算器型号推荐购买列表，工欲善其事必先利其器，买个官方认可的计算器带上考场，用起来也更得心应手。

目 录

第一部分　文字题：结构分析骨干提取法 ……………………………………… 1
　Part 1　结构分析骨干提取法典型题精讲 ………………………………………… 3
　Part 2　结构分析骨干提取法典型题精练 ………………………………………… 16

第二部分　易错题分析 …………………………………………………………… 25

第三部分　简单粗暴解析法 ……………………………………………………… 33

第四部分　数学必做 100 题 ……………………………………………………… 49
　Practice 1 ……………………………………………………………………………… 51
　Practice 1　答案与解析 ……………………………………………………………… 54
　Practice 2 ……………………………………………………………………………… 58
　Practice 2　答案与解析 ……………………………………………………………… 62
　Practice 3 ……………………………………………………………………………… 66
　Practice 3　答案与解析 ……………………………………………………………… 70
　Practice 4 ……………………………………………………………………………… 74
　Practice 4　答案与解析 ……………………………………………………………… 78
　Practice 5 ……………………………………………………………………………… 83
　Practice 5　答案与解析 ……………………………………………………………… 87

第五部分　全真模考题 …………………………………………………………… 91
　Test 1 ………………………………………………………………………………… 93
　Test 1　答案与解析 ………………………………………………………………… 109
　Test 2 ………………………………………………………………………………… 118
　Test 2　答案与解析 ………………………………………………………………… 134

第六部分　考前 5 天复习方案 ………………………………………………… 145
　Day 1 ………………………………………………………………………………… 147
　Day 2 ………………………………………………………………………………… 151
　Day 3 ………………………………………………………………………………… 155
　Day 4 ………………………………………………………………………………… 160
　Day 5 ………………………………………………………………………………… 164

　附录 1　知识点清单 ………………………………………………………………… 165

1. 数与数的运算 ………………………………………………………… 165
2. 因子与倍数 …………………………………………………………… 167
3. 比例 …………………………………………………………………… 168
4. 百分比 ………………………………………………………………… 170
5. 统计与概率 …………………………………………………………… 173
6. 乘方与根数 …………………………………………………………… 176
7. 代数 …………………………………………………………………… 177
8. 解析几何 ……………………………………………………………… 180
9. 函数与数学模型 ……………………………………………………… 183
10. 平面几何 …………………………………………………………… 186
11. 三角函数 …………………………………………………………… 192
12. 立体几何 …………………………………………………………… 196

附录2 常见数学表达 …………………………………………………… 200
1. 基本概念和词组 ……………………………………………………… 200
2. 运算 …………………………………………………………………… 201
3. 比率和比例 …………………………………………………………… 202
4. 等式和不等式 ………………………………………………………… 202
5. 函数 …………………………………………………………………… 203
6. 线和角 ………………………………………………………………… 203
7. 三角形 ………………………………………………………………… 204
8. 四边形和多边形 ……………………………………………………… 205
9. 圆 ……………………………………………………………………… 206
10. 立体几何 …………………………………………………………… 206
11. 坐标几何 …………………………………………………………… 207
12. 概率 ………………………………………………………………… 207
13. 集合与数列 ………………………………………………………… 208
14. 数据分析 …………………………………………………………… 208
15. 单位 ………………………………………………………………… 208
16. 常见美制单位换算 ………………………………………………… 209

附录3 数学计算器型号推荐购买列表 ………………………………… 210

致谢 ……………………………………………………………………… 211

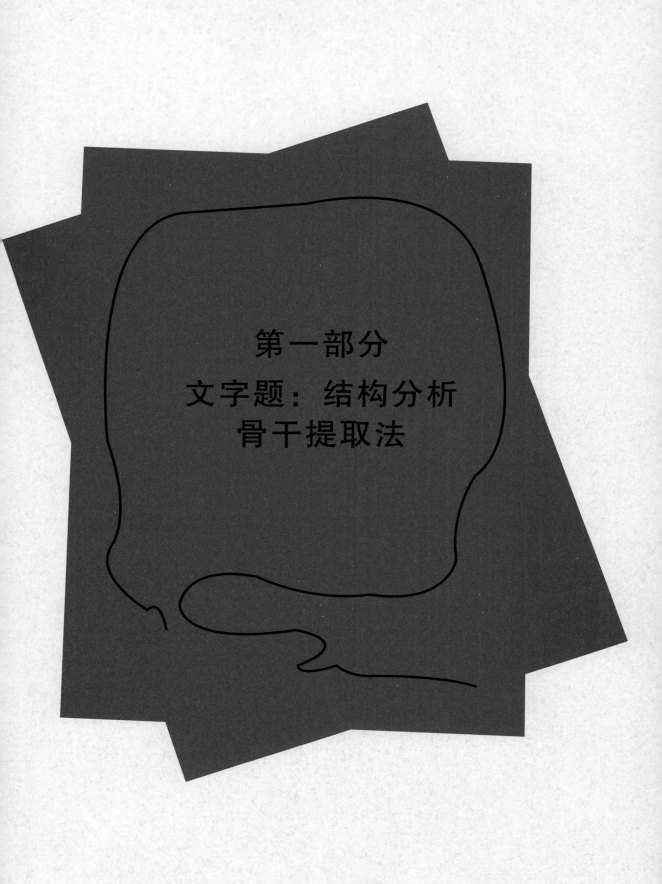

第一部分

文字题：结构分析
骨干提取法

Part 1　结构分析骨干提取法典型题精讲

The flu shot for a flu season is created from four strains of the flu virus, named Strain A, B, C and D, respectively. Medical researchers use the following data to determine the effectiveness of the vaccine over the flu season. Table 1 - 1 shows the effectiveness of the vaccine against each of these strains individually. The graph (Fig. 1 - 1) below the table shows the prevalence of each of these strains during each month of the flu season, represented as a percentage of the overall cases of flu that month.

Table 1 - 1

Strain	Effectiveness
A	35%
B	13%
C	76%
D	68%

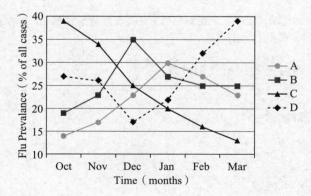

Fig. 1 - 1

For the strain against which the flu shot was the most effective, approximately how effective was the shot overall when that strain was least prevalent?

A. 13%　　　　　　　　　　　　B. 20%
C. 27%　　　　　　　　　　　　D. 48%

SAT-1 数学轻松突破 800 分：思路与技巧的飞跃

■分步详解

结构分析：新 SAT-1 数学的文字题出题特征为"三部论"：第一部分为背景知识，第二部分为条件，第三部分为问题与作答。

针对三部分我们的策略是：略读背景知识 + 条件判断 + 精读问题

【1】The flu shot for a flu season is created from four strains of the flu virus, named Strain A, B, C and D, respectively.	【T1】文字题首句功能：介绍背景。关键词：flu shot, Strain A, B, C and D ↓
【2】Medical researchers use the following data to determine the effectiveness of the vaccine over the flu season. Table 1-1 shows the effectiveness of the vaccine against each of these strains individually. The graph below the table shows the prevalence of each of these strains during each month of the flu season, represented as a percentage of the overall cases of flu that month.	【2】给出条件，数据在表 1-1 和图中 此题考查百分比的知识点，其中表 1-1 数据之和不为 1，而图中提到了 % of all case，由此判断此题考查的核心知识范畴为：权重系数 $X_1 \times f_1 + X_2 \times f_2 + X_3 \times f_3 + \cdots + X_n \times f_n$ ↓
【3】For the strain against which the flu shot was the most effective, approximately how effective was the shot overall when that strain was least prevalent?	【3】问题一般结构是"问题 + 条件"，本问的句子比较长，也是此题的难点，其中需要有一些句子分析，提取关键词： (A)问题是：how effective was the shot overall (B)条件：分别由一个 which 和一个 when 引导的从句给出：the strain against 是 the most effective + least prevalent，因此我们找这样一个菌株 (strain)，要满足疫苗对它的作用最好，而且要在该菌株最不流行的时候。

解析	
Step 1 找条件，这样一个菌株 (strain)，要满足疫苗对它的作用最好，而且要在该菌株最不流行的时候。 **Step 2** 读表发现对 Strain C 而言疫苗作用效果最好，为 76%。 ↓ 读图发现 Strain C 在 Mar 最不流行，约为 13% of all cases。 ↓ **Step 3** 根据此题的知识点计算：在 Mar： Strain A = 23% × 35% Strain B = 25% × 13% Strain C = 13% × 76% Strain D = 39% × 68% A + B + C + D ≈ 0.477 ≈ 48%。	条件为原题第二部分，即在表和图中

答案	D

第一部分 文字题：结构分析骨干提取法

Questions 2 and 3 refer to the following information.
Professor Malingowski, a chemist and teacher at a community college, is organizing his graduated cylinders in the hopes of keeping his office tidy and setting a good example for his students. He has beakers with diameters, in inches, of $\frac{1}{2}$, $\frac{3}{4}$, $\frac{4}{5}$, 1 and $\frac{5}{4}$.

Professor Malingowski notices one additional cylinder lying on the ground, and can recall certain facts about it, but not its actual diameter. If he knows that the value of the additional graduated cylinder's diameter, x, will not create any modes and will make the mean of the set equal to $\frac{5}{6}$, what is the value of the additional cylinder's diameter?

With his original five cylinders, Professor Malingowski realizes that he is missing a cylinder necessary for his upcoming lab demonstration for Thursday's class. He remembers that the cylinder he needs, when added to the original five, will create a median diameter value of $\frac{9}{10}$ for the set of six total cylinders. He also knows that the measure of the sixth cylinder will exceed the value of the range of the current five cylinders by a width of anywhere from $\frac{1}{5}$ inches to $\frac{1}{2}$ inches, inclusive. Based on the above data, what is one possible value of y, the diameter of this missing sixth cylinder?

■分步详解

结构分析：此题为两个问题的套题。从整体上看还是分成三个部分：第一部分为总体背景知识，每一个问题又分为背景知识、条件以及问题与作答。
针对三部分我们的策略是：**略读背景知识 + 条件判断 + 精读问题**

第一问： 【1.1】Professor Malingowski, a chemist and teacher at a community college, is organizing his graduated cylinders in the hopes of keeping his office tidy and setting a good example for his students. He has beakers with diameters, in inches, of $\frac{1}{2}$, $\frac{3}{4}$, $\frac{4}{5}$, 1 and $\frac{5}{4}$.	【1.1】文字题首句功能：介绍背景。关键词：graduated cylinders（刻度容器），$\frac{1}{2}$, $\frac{3}{4}$, $\frac{4}{5}$, 1 and $\frac{5}{4}$ ↓

【1.2】 Professor Malingowski notices one additional cylinder lying on the ground, and can recall certain facts about it, but not its actual diameter. If he knows that the value of the additional graduated cylinder's diameter, x, will not create any modes and will make the mean of the set equal to $\frac{5}{6}$,	【1.2】问题1的补充条件，增加了一个 cylinder，diameter 为 x，没有增加 mode，平均值为 $\frac{5}{6}$。 ↓ 此题给出平均值、众数等条件，由此判断考查的核心知识范畴为：众数和平均值等概念（详见本书附录1）。 ↓
【1.3】 what is the value of the additional cylinder's diameter?	【1.3】问题一般结构是：问题＋条件，本问比较简单，求增加的那个 cylinder 的 diameter 是多少。

解析	
Step 1 找条件：原来有五个值，现在增加了一个值，众数没有变，新的平均值给出了。 ↓ **Step 2** 根据此题的知识点计算： $$\frac{5}{6} = \frac{\frac{1}{2}+\frac{3}{4}+\frac{4}{5}+1+\frac{5}{4}+x}{6}$$ 得到 $x = \frac{7}{10}$ 或者 0.7。	众数没有变化，说明新增的值不在上述已有的五个值之中。

答案	$\frac{7}{10}$ 或者 0.7

第二问： 【2.1】With his original five cylinders, Professor Malingowski realizes that he is missing a cylinder necessary for his upcoming lab demonstration for Thursday's class.	【2.1】文字题首句功能：介绍背景，介绍说这个教授遗漏了一个 cylinder，可以略读。 ↓
【2.2】He remembers that the cylinder he needs, when added to the original five, will create a median diameter value of $\frac{9}{10}$ for the set of six total cylinders. He also knows that the measure of the sixth cylinder will exceed the value of the range of the current five cylinders by a width of anywhere from $\frac{1}{5}$ inches to $\frac{1}{2}$ inches, inclusive.	【2.2】问题的两个补充条件： 增加了一个 cylinder 后得到了一个中位数（median diameter value）是 $\frac{9}{10}$ ↓ 此题考查中位数的概念，由于一共是六个值，中位数是按照从小到大排序后的第三、四个值的算术平均值 ↓ 增加的这个 cylinder 直径超过（exceed）了现有五个值范围（range）的 $\frac{1}{5}$ 到 $\frac{1}{2}$ 之间，且包含（inclusive） ↓

第一部分 文字题：结构分析骨干提取法

【2.3】Based on the above data, what is one possible value of y, the diameter of this missing sixth cylinder?	【2.3】问第六个值的一个可能值，说明这个值不止一个。 答案不唯一，一般是给出一个范围。
解析	
Step 1 此题考查中位数的计算方法，新的中位数为 $\frac{9}{10}$，正好满足 $\frac{1}{2} \times \left(\frac{4}{5} + 1\right) = \frac{9}{10}$，由此判断 $\frac{4}{5}$ 和 1 是第三、四个值，说明新增的第六个值 $\geqslant 1$。 ↓ **Step 2** 原有五个值的范围（range）为 $\frac{5}{4} - \frac{1}{2} = \frac{3}{4}$ ↓ 现有第六个值 y 应该在 $\frac{3}{4} + \frac{1}{5} = \frac{19}{20}$ 和 $\frac{3}{4} + \frac{1}{2} = \frac{5}{4}$ 之间。 ↓ **Step 3** 满足以上两个条件，故 $1 \leqslant y \leqslant \frac{5}{4}$，此题问的是一个可能值，答案只要满足 $1 \leqslant y \leqslant \frac{5}{4}$ 均可。	中位数的计算需要注意：由于数字大小不一，当插入一个数字后，很可能原有的顺序会发生变化，例如在本题中： $$\frac{1}{2},\ \frac{3}{4},\ \frac{4}{5},\ 1,\ \frac{5}{4}$$ (A) 如果新增的值 $y \geqslant 1$，则为： $$\frac{1}{2},\ \frac{3}{4},\ \frac{4}{5},\ 1,\ y,\ \frac{5}{4}$$ 因此中位数为 $\frac{4}{5}$ 和 1 的算术平均值 (B) 如果新增的值 $\frac{4}{5} \leqslant y \leqslant 1$，则为： $$\frac{1}{2},\ \frac{3}{4},\ \frac{4}{5},\ y,\ 1,\ \frac{5}{4}$$ 中位数为 $\frac{4}{5}$ 和 y 的算术平均值 可见，插入重排的位置不同，中位数的计算值也相应有不同。
答案	$1 \leqslant y \leqslant \frac{5}{4}$ 范围内的任意值

Questions 4 and 5 refer to the following information.

The table below (Table 1-2) shows the population distribution for the 2,400 occupants of the city of Centre Hill.

Table 1-2

	Adult Male	Adult Female	Child
Living in Uptown(%)	9	8	6
Living in Midtown(%)	22	20	15
Living in Downtown(%)	21	22	12
Living in Suburbs(%)	48	50	67

If there are an equal number of adults and children, and adult females outnumber adult males by 200, what is the sum of the women living uptown and the children living in the suburbs of Centre Hill?

SAT-1 数学轻松突破 800 分：思路与技巧的飞跃

Centre Hill plans to annex the area around a nearby lake. This new part of Centre Hill will be called, appropriately, The Annex. The Annex will add to the current population of Centre Hill. The percent of adult males living in Uptown will decrease to 6% after incorporating The Annex into Centre Hill. If the information from question 4 holds true for the original four districts of the city of Centre Hill, then how many adult males live in The Annex?

■ 分步详解

结构分析：新 SAT-1 数学的文字题出题特征为"三部论"，第一部分为背景知识，第二部分为条件，第三部分为问题与作答。
针对三部分我们的策略是：略读背景知识 + 条件判断 + 精读问题

背景知识：	本题的背景知识或内容不是很多，主要信息出现在表格中。
【T1】The table below（Table 1-3）shows the population distribution for the 2,400 occupants of the city of Centre Hill.	【T1】总人口为 2,400，表格中给出四个区域，统计的人数分类是女性成人、男性成人和孩子。

Table 1-3

	Adult Male	Adult Female	Child
Living in Uptown(%)	9	8	6
Living in Midtown(%)	22	20	15
Living in Downtown(%)	21	22	12
Living in Suburbs(%)	48	50	67

第一问：	
【1.1】If there are an equal number of adults and children, and adult females outnumber adult males by 200，【1.2】what is the sum of the women living uptown and the children living in the suburbs of Centre Hill?	【1.1】本题的主要背景已经给出，第一题的第一句话直接给出本题的条件： （A）孩子与成人的数量相等 （B）女性比男性多（outnumber）200 人 【1.2】问题一般结构是：问题 + 条件，提取关键词： （A）问题是：求人数的总和 （B）条件：住在上城区的女性 + 住在郊区的孩子

第一部分　文字题：结构分析骨干提取法

解析	
Step 1 根据背景条件用首字母做设。设男性成人的数量为 m，女性成人的数量为 f，由于总人口数为 2,400（已经在背景中给出）+ 孩子与成人的数量相等（本题条件给出）+ 女性比男性多 200 人（本题条件给出），因此可以得到以下关系： 孩子和成人均为总人口的一半，即 1,200 人 $f + m = 1,200$ $f - m = 200$ 两个式子联立可以解出 $f = 700$，$m = 500$。 ↓ **Step 2** 问题要求：住在上城区的女性和住在郊区的孩子数量之和，因此可得： 住在上城区的女性数量：$700 \times 8\% = 56$； 住在郊区的孩子数量：$1,200 \times 67\% = 804$。 ↓ **Step 3** 根据题意，求上述两个部分的总和：$56 + 804 = 860$。	需要结合背景条件和本题的条件。

答案　860

第二问： 【2.1】Centre Hill plans to annex the area around a nearby lake. This new part of Centre Hill will be called, appropriately, The Annex. The Annex will add to the current population of Centre Hill. 【2.2】The percent of adult males living in Uptown will decrease to 6% after incorporating The Annex into Centre Hill. 【2.3】If the information from question 4 holds true for the original four districts of the city of Centre Hill, then how many adult males live in The Annex?	【2.1】本题的背景：介绍新增了一个区域的情况，可以略读。 ↓ 【2.2】本题的条件：在新增一个区域后，住在上城区的男性比例发生了变化。 ↓ 【2.3】如果第四题的数据没有问题的话，问有多少男性成人住在新的区域？ ↓ 本题是比例题的另一种典型考法，即考虑在加入某些元素之后，百分比发生了变化，特别是百分比下降。百分比的本质是比率，引起这个变化的一个原因是分母的变化（在分子组分不变的情况下，分母组分变大，最后的比值就会变小） ↓ 解决本题的关键是找比值变化中的不变组分

解析	
Step 1 根据题意，比值变化中的不变组分为住在上城区的男性数量，题中又提示"如果第四题的数据没有问题"，这说明这个数据需要从第四题的答案中获得的。 在第四题中得到了 $m = 500$，即男性成人总人数为 500 人。从背景条件中可以得到上城区的男性比例为 9%，因此可得住在上城区的男性数量为： $500 \times 9\% = 45$ 人。 ↓ **Step 2** 从本题的条件"在增加一个新的区域后，住在上城区的男性比例下降到了 6%"，可以反求新的男性成人总人数（T_m）： $45 = T_m \times 6\%$，得到 T_m 为 750。 ↓ **Step 3** 变化前男性成人总人数为 500 人，现在男性成人总人数为 750 人，之间的差值 $750 - 500 = 250$ 人即是住在新区域中的男性成人总人数。	本题的比值是住在上城区的男性数量比男性成人总人数，因此需要找到这对数据之间的变与不变关系。 下降的原因：住在上城区的男性数量不变，而男性成人总人数在新增一个区域后发生了变化。
答案	250

Questions 6 and 7 refer to the following information.

A company sponsors a health program for its employees by partnering with a local gym. If employees pay for a yearlong membership at this gym, then for every day the employee uses his or her swipe card to enter the gym (and work out), the company reimburses the employee 0.2% of the cost of the $220 membership. Additionally, any employee who goes to the gym more than 60% of the days in the year gets one bonus paid day off of work. The company uses a 365-day year.

If 246 employees participate in the program and they each go to the gym an average of 84 days per year, how much money in membership reimbursements will the company pay out? Round your answer to the nearest whole dollar.

Giving employees additional paid time off also costs the company money because it is paying the salary of an employee who is not actually doing any work on that day. The pie graph (Fig. 1-2) below shows gym usage for the 246 employees who participated in the health program.

第一部分　文字题：结构分析骨干提取法

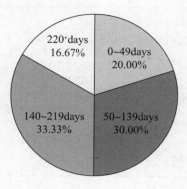

Fig. 1-2

If the average salary of workers who participated was $14.90 per hour and one day off equals 8 hours, how much did the health program day-off benefit cost the company? Round your answer to the nearest whole dollar.

■ 分步详解

结构分析：新 SAT-1 数学的文字题出题特征为"三部论"，第一部分为背景知识，第二部分为条件，第三部分为问题与作答。
　　针对三部分我们的策略是：**略读背景知识＋条件判断＋精读问题**

背景知识： 【T1】A company sponsors a health program for its employees by partnering with a local gym. 【T2】If employees pay for a yearlong membership at this gym, then for every day the employee uses his or her swipe card to enter the gym（and work out）, the company reimburses the employee 0.2% of the cost of the $220 membership. 【T3】Additionally, any employee who goes to the gym more than 60% of the days in the year gets one bonus paid day off of work. 【T4】The company uses a 365-day year.	本题的主要背景在一开始已经给出： 【T1】介绍本题的前提与背景，可以略读。 ↓ 【T2】给出本题的第一个条件：公司给参加运动的员工报销（reimburse），按每天（every day）花销的 0.2% 支付。 ↓ 【T3】对运动量超过总量 60% 的员工可以有一个额外津贴（bonus）。 ↓ 【T4】附属条件：一年按 365 天计算，呼应【T2】中的一年期的会员资格（a yearlong membership）。
第一问： 【1.1】If 246 employees participate in the program and they each go to the gym an average of 84 days per year, 【1.2】how much money in membership reimbursements will the company pay out? 【1.3】Round your answer to the nearest whole dollar.	【1.1】本题的第一个条件，给出两个数据：246 和 84。 【1.2】给出本题问题，问公司要支付多少钱？需要注意：上述题目给出的条件，说明这里的计算是需要经过筛选，找出符合条件的数据。 【1.3】对最终的结果还要求保留到整数（这是很多同学容易忽略的地方）。
解析	

Step 1 根据题目的条件，公司对参加运动的员工每天报销额度为 $ 220×0.2% = $ 0.44。 **Step 2** 筛选符合条件的数据：员工数：246，天数：84。 ↓ **Step 3** 计算公司需要支付的费用： $ 0.44×84×246 = $ 9,092.16 ≈ $ 9,092。	这道题的本质是"单位量×数量"的问题，在新 SAT-1 数学中经常出现的以下几组关系，需要同学们注意： 单价×数量 速度×时间 单位组分×组数
答案　　9,092	
第二问： 【2.1】Giving employees additional paid time off also costs the company money because it is paying the salary of an employee who is not actually doing any work on that day. 【2.2】The pie graph below shows gym usage for the 246 employees who participated in the health program. 【2.3】If the average salary of workers who participated was $ 14.90 per hour and one day off equals 8 hours, 【2.4】how much did the health program day-off benefit cost the company? 【2.5】Round your answer to the nearest whole dollar.	本题在原题的基础上又增加了新的条件。 【2.1】为本题的新增条件，主要引出公司成本（cost）的概念，这里的"增加支付部分"（additional paid）与题目的条件【T1】呼应。 ↓ 【2.2】以图的形式给出本题的条件：总人数为 246 人，并从饼状图中可以看出是按照天数进行分组的。 【2.3】给出本题的第 2 个条件： (A) 每天的工资为 $ 14.90； (B) 每天工作 8 小时。 ↓ 【2.4】本题的问题：公司在此项目中的成本为多少？ 【2.5】对最终的结果还要求保留到整数。
解析	
Step 1 筛选符合条件的数据： 条件【T3】中指出公司只针对一年运动天数超过 60% 的员工额外津贴，以及【T4】中指出一年总天数为 365 天。因此，首先计算满足的条件组，即：365×60% = 219 天，要在图中选择大于 219 天的那一组或几组。通过读图后发现符合条件的为 "220⁺ day" 组。 ↓ **Step 2** 根据上述数据，得到满足条件的人数为： 246×16.67% ≈ 41 人。 ↓ **Step 3** 计算公司需要支付的费用： 每小时：$ 14.90 每人每天工作时间：8 小时 人数：41 人 费用为 14.90×8×41 = 4,887.2 ≈ 4,887。	这道题的本质是"单位量×数量"的问题，因此需要： (A) 筛选符合条件的（对应图中某一类数据）； (B) 找到单位量和数量。
答案　　4,887	

第一部分 文字题：结构分析骨干提取法

Questions 8 and 9 refer to the following information.

Mercury is a naturally occurring metal that can be harmful to humans. The current recommendation is for humans to take in no more than 0.1 microgram for every kilogram of their weight per day. Fish generally carry high levels of mercury, although certain fish have higher mercury content than others. Fish, however, are healthy sources of many other nutrients, so nutritionists recommend keeping them in the human diet. The figure (Fig. 1-3) below shows the average mercury content of several types of fish.

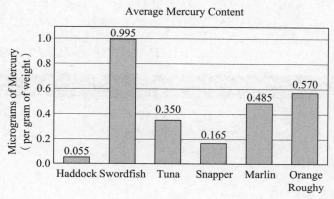

Fig. 1-3

If a person weighs 82 kilograms, how many grams of snapper can he safely consume per day? Round your answer to the nearest gram.

Suppose in a week: a person regularly eats one portion of each type of the fish shown in the bar graph, except the fish with the highest mercury content. What is this person's average daily mercury consumption, in micrograms, assuming a portion size of 100 grams? Round your answer to the nearest microgram.

■ 分步详解

结构分析：新SAT-1数学的文字题出题特征为"三部论"，第一部分为背景知识，第二部分为条件，第三部分为问题与作答。
针对三部分我们的策略是：**略读背景知识 + 条件判断 + 精读问题**

背景知识：	本题的背景知识比较多【T1-T4】，但是根据出
【T1】Mercury is a naturally occurring metal that can be harmful to humans. 【T2】The current recommendation is for humans to take in no more than 0.1 microgram for every kilogram of their weight per day. 【T3】Fish generally carry high levels of mercury, although certain fish have higher mercury content than others. Fish, however, are healthy sources of many other nutrients, so nutritionists recommend keeping them in the human diet.	题的基本思路，这些信息对解题关系不大，可以略读。主要数据从图中读出即可。

【T4】 The figure below shows the average mercury content of several types of fish.

第一问：

【1.1】 If a person weighs 82 kilograms, 【1.2】 how many grams of snapper can he safely consume per day? 【1.3】 Round your answer to the nearest gram.

【1.1】本题第一问的条件比较简单，仅根据此条件无法解题，所以需要在原题的条件中找到相应关系。本题中提到"人的体重"正好呼应【T2】。
↓
【1.2】为本题的问题，问他每天摄入多少克的鲷鱼是安全的？
通过"鲷鱼（snapper）"可以对应在图中找到相应的数据。本题的关键是汞（mercury）的含量。因此可以通过这些数据建立起"人的体重"和"鱼的重量"之间的关联。
【1.3】对最终的结果还要求保留到整数。

解析

Step 1
根据题意，结合【T2】中的条件，人体汞摄入量不超过的量为 $0.1 \times 82 = 8.2$ 毫克。
↓
Step 2
根据图找到与"鲷鱼"相关的数据，设每天摄入鲷鱼为 x 克，可得：
$x \times 0.165 \leq 8.2$，解得 $x \leq 49.697 \approx 49$（克）。

注意由于原题要求保留到整数，所以不能用四舍五入。

答案 49

第二问：

【2.1】 Suppose in a week: a person regularly eats one portion of each type of the fish shown in the bar graph, except the fish with the highest mercury content. 【2.2】 What is this person's average daily mercury consumption, in micrograms, assuming a portion size of 100 grams? 【2.3】 Round your answer to the nearest microgram.

【2.1】为本题新增的条件，需要注意到两个表述：
（A）排除（except）
（B）最高（highest）
一般在新 SAT-1 数学题的条件表述中需要注意到特殊值：如最值（最大值和最小值）以及排除（except）等语句。
↓
【2.2】本题的问题：平均每天汞的摄入量为多少？（以毫克来计算，假设每条鱼摄入量为 100 克）
需要注意到：
（A）【2.1】中提到的每一种（each type）。因此本题中的数据要包括所有鱼，但是要排除汞含量最高的那种鱼。
（B）条件中给出的是周（week）的数据，而问题中问的是"平均每天摄入量（average daily mercury consumption）"。因此这里面还需要换算。
【2.3】对最终的结果还要求保留到整数。

第一部分 文字题：结构分析骨干提取法

解析	
Step 1 从条件中找到每一种鱼的数据，但是要排除掉汞含量最高的那种鱼：Swordfish。因此计算得到： Haddock：$0.055 \times 100 = 5.5$ 毫克； Swordfish：汞含量最高，排除； Tuna：$0.350 \times 100 = 35$ 毫克； Snapper：$0.165 \times 100 = 16.5$ 毫克； Marlin：$0.485 \times 100 = 48.5$ 毫克； Orange Roughy：$0.570 \times 100 = 57$ 毫克。 ↓ **Step 2** 因此这一周通过摄入鱼肉而摄入的汞含量为： $5.5 + 35 + 16.5 + 48.5 + 57 = 162.5$（毫克）。 ↓ **Step 3** 平均每天的摄入量为： $\frac{162.5}{7} = 23.214 \approx 23$（毫克）。	本题的计算中需要注意到时间的换算，这也是新SAT-1数学常考的内容。由于题目中采用 week 和 average daily 来表示时间关系，使得学生在计算的时候往往会忽视两者之间的换算。
答案 23	

Part 2 结构分析骨干提取法典型题精练

A museum is building a scale model of Sue, the largest *Tyrannosaurus rex* skeleton ever found. Sue was 13 feet tall and 40 feet long, and her skull had a length of 5 feet. If the length of the museum's scale model skull is 3 feet and 1.5 inches, what is the difference between the scale model's length and its height?

A. 8 feet and 1.5 inches
B. 16 feet and 10.5 inches
C. 25 feet
D. 27 feet and 4 inches

A power company divides the geographic regions it serves into grids. The company is able to allocate the power it generates based on the usage and needs of a particular grid. Certain grids use more power at certain times of the day, so companies often shift power around to different grids at various times. On any given day, the company makes several changes in the power allocation to Grid l. First, it increases the power by 20%. Then, it decreases it by 10%. Finally, it increases it by 30%. What is the net percent increase in this grid's power allocation? (Round to the nearest whole percent and ignore the percent sign when entering your answer.)

Questions 3 and 4 refer to the following information.
Every Saturday morning, three friends meet for breakfast at 9:00 AM. Andrea walks, Kellan bikes, and Joelle drives.

Last Saturday, all three friends were exactly on time. Andrea left her house at 8:30 AM and walked at a rate of 3 mph. Kellan left his house at 8:15 AM and biked at a rate of 14 mph. Joelle left her house at 8:45 AM and drove an average speed of 35 mph. How many miles from the restaurant does the person who travelled the farthest live?

Kellan lives 12 miles away from Andrea. On a different Saturday, Kellan biked at a rate of 15 mph to Andrea's house. The two then walked to the restaurant at a rate of 2.5 mph, and they arrived five minutes early. What time did Kellan leave his house? Enter your answer as three digits and ignore the colon. For example, if your answer is 5:30

AM, enter 530.

Questions 5 and 6 refer to the following information.

Bridget is starting a tutoring business to help adults get their GEDs. She already has five clients and decides they can share a single textbook, which will be kept at her office, and that she also needs one notebook and four pencils for each of them. She records her supply budget, which includes tax, in the table (Table 1-4) shown.

Table 1-4

Supply	Total Number Needed	Cost Each
Textbook	1	$24.99
Notebooks	5	$3.78
Pencils	20	$0.55

The textbook makes up what percent of Bridget's total supply budget? Round to the nearest tenth of a percent and ignore the percent sign when entering your answer.

Bridget's business does very well, and she needs more supplies. She always orders them according to the table above, for five clients at a time. At the beginning of this year, she orders the supplies for the whole year, which cost $988.02. Halfway through the year, she decides to take inventory of the supplies. She has used $713.57 worth of the supplies. How many pencils should be left, assuming the supplies were used at the rate for which she originally planned?

Questions 7 and 8 refer to the following information.

The amount of glucose, or sugar, in a person's blood is the primary indicator of diabetes. When a person fasts (doesn't eat) for eight hours prior to taking a blood sugar test, his/her glucose level should be below 100 mg/dL. A person is considered at risk for diabetes, but is not diagnosed as diabetic, when fasting glucose levels are between 100 and 125. If the level is above 125, the person is considered to have diabetes. The table (Table 1-5) below shows the ages and glucose levels of a group of diabetes study participants.

Table 1-5   Diabetes Study Results

Age Group	<100 mg/dL	100~125 mg/dL	>125 mg/dL	Total
18~25	9	22	17	48
26~35	16	48	34	98
36~45	19	35	40	94
older than 45	12	27	21	60
Total	56	132	112	300

According to the data, which age group had the smallest percentage of people with a healthy blood sugar level?

A. 18~25　　　B. 26~35　　　C. 36~45　　　D. older than 45

Based on the table, if a single participant is selected at random from all the participants, what is the probability that he or she will be at risk for diabetes and be at least 36 years old?

A. $\dfrac{7}{60}$　　　B. $\dfrac{11}{25}$　　　C. $\dfrac{31}{77}$　　　D. $\dfrac{31}{150}$

Questions 9 and 10 refer to the following information.

Eli left his home in New York and traveled to Brazil on business. On Monday, he used his credit card to purchase these pewter vases, R$ (Brazilian reais) 128, R$ 66 and R$ 85, as a gift for his wife. For daily purchases totaling less than 200 U.S. dollars, Eli's credit card company charges a 2% fee. If the total charge on his credit card for the vases was $126.48, and no other purchases were made, what was the foreign exchange rate on Monday in Brazilian reais per U.S. dollar? If necessary, round your answer to the nearest hundredth.

On Wednesday, Eli bought a tourmaline ring that cost R$ 763. For daily purchases over $200, his credit card company charges the same 2% fee on the first $200 of the converted price and 3% on the portion of the converted price that is over $200. If the total charge on his credit card for the ring was $358.50, what was the amount of decrease, as a percentage, in the foreign exchange rate between Monday and Wednesday? Round your answer to the nearest whole percent.

■ 答案与解析

1	B	【结构解析】： 【1】A museum is building a scale model of Sue, the largest *Tyrannosaurus rex* skeleton ever found. 【2】Sue was 13 feet tall and 40 feet long, and her skull had a length of 5 feet. 【3】If the length of the museum's scale model skull is 3 feet and 1.5 inches, 【4】what is the difference between the scale model's length and its height? 【1】：背景，略读。【2-3】：条件。【4】：问题。 【题干】： 一家博物馆为迄今为止发现的最大尺寸霸王龙 Sue 按比例做了一副模型。Sue 高 13 英尺，长 40 英尺，头盖骨长 5 英尺，如果 Sue 在博物馆的等比例模型其头盖骨长为 3 英尺 1.5 英寸。

第一部分　文字题：结构分析骨干提取法

1	B	**【问题】：** Sue 的模型其长与高的差值是多少？ **【解析】：** 根据题意可得，本题的核心考点是实物与模型尺寸之间的比例及其换算。由于本题中出现了两个尺寸单位（英尺与英寸），所以需要对其进行转化使得单位统一。 **Step 1** 头盖骨：实物长为 $5\times12=60$ 英寸，模型长为 $3\times12+1.5=37.5$ 英寸。因此可以得到实物与模型之间的比例为 $\frac{60}{37.5}=\frac{8}{5}$。 **Step 2** 根据该比例以及实物的尺寸，可以计算出模型的尺寸，即： 模型的长度：$\frac{40}{\frac{8}{5}}=25$ 英尺，模型的高度：$\frac{13}{\frac{8}{5}}=8.125$ 英尺。 **Step 3** 最后计算模型长与高的差值为：25 英尺 -8.125 英尺 $=16.875$ 英尺 $=16$ 英尺 10.5 英寸，选 B。
2	40	**【结构解析】：** 【1】A power company divides the geographic regions it serves into grids. 【2】The company is able to allocate the power it generates based on the usage and needs of a particular grid. 【3】Certain grids use more power at certain times of the day, so companies often shift power around to different grids at various times. 【4】On any given day, the company makes several changes in the power allocation to Grid l. 【5】First, it increases the power by 20%. 【6】Then, it decreases it by 10%. 【7】Finally, it increases it by 30%. 【8】What is the net percent increase in this grid's power allocation? (【9】Round to the nearest whole percent and ignore the percent sign when entering your answer.) 【1-4】：本题总背景，可以略读。【5-7】：条件。【8-9】：问题与要求。 **【题干】：** 一家电力公司将所服务的区域进行网格划分。这样基于特定网格的能源使用和需求情况，该公司可以进行电力分配。有些网格在一天的某些特定时间需求量会更多。因此该电力公司在不同时间会调配不同电力去不同的地方。该公司某一天在网格 1 区域内多次调配不同的电力。首先使电力增加了 20%，之后下降了 10%，最后增加了 30%。 **【问题】：** 该网格区域的电力净增加的百分比是多少？计算结果最后保留到整数，并且在填涂时忽略符号"%"。 **【解析】：** 根据题意可得，本题考查有关百分比的计算。这类题由于没有给出具体的数值，可以考虑设起始值为 100，以便于计算： 第一次变化：$100\times(1+20\%)=120$； 第二次变化：$120\times(1-10\%)=108$； 第三次变化：$108\times(1+30\%)=140.4$； 因此，百分比净增加为 $\frac{140.4-100}{100}\times100\%=40.4\%\approx40\%$。

3	10.5	【结构解析】： 【1】Every Saturday morning, three friends meet for breakfast at 9:00 AM. Andrea walks, Kellan bikes, and Joelle drives. 【2】Last Saturday, all three friends were exactly on time. 【3】Andrea left her house at 8:30 AM and walked at a rate of 3 mph. 【4】Kellan left his house at 8:15 AM and biked at a rate of 14 mph. 【5】Joelle left her house at 8:45 AM and drove an average speed of 35 mph. 【6】How many miles from the restaurant does the person who travelled the farthest live? 【1】：本题总背景，可以略读，但是需要注意"9点碰头"这个条件。【2】：本题背景，可以略读。【3-5】：条件。【6】：问题。 【题干】： 在每个周六上午，三位朋友相约 9:00 一起吃早饭。其中 Andrea 步行，Kellan 骑车，而 Joelle 开车。上个周六，三位朋友又一次相约，Andrea 在 8:30 离开家，步行的速度是 3 英里/小时，Kellan 在 8:15 离开家，骑车的速度为 14 英里/小时，Joelle 在 8:45 离开家，开车的速度为 35 英里/小时。 【问题】： 离饭店最远的那个人移动的距离是多少？ 【解析】： 根据题意可得，本题考查的关键：速度与时间的对应关系以及不同时间单位的转换。 Andrea：从 8:30 离开家至 9:00 到饭店，用时 30 分钟即 0.5 小时，距离为 $3 \times 0.5 = 1.5$ 英里； Kellan：从 8:15 离开家至 9:00 到饭店，用时 45 分钟即 0.75 小时，距离为 $14 \times 0.75 = 10.5$ 英里； Joelle：从 8:45 离开家至 9:00 到饭店，用时 15 分钟即 0.25 小时，距离为 $35 \times 0.25 = 8.75$ 英里； 因此离饭店最远的人是 Kellan，距离为 10.5 英里。
4	731	【结构解析】： 【1】Kellan lives 12 miles away from Andrea. 【2】On a different Saturday, Kellan biked at a rate of 15 mph to Andrea's house. 【3】The two then walked to the restaurant at a rate of 2.5 mph, and they arrived five minutes early. 【4】What time did Kellan leave his house? 【5】Enter your answer as three digits and ignore the colon. For example, if your answer is 5:30 AM, enter 530. 【1】：本题背景，可以略读。【2-3】：条件。【4-5】：问题与要求。 【题干】： Kellan 居住的地方离 Andrea 有 12 英里远，在另一个周六，Kellan 以 15 英里/小时的速度骑车去 Andrea 家，之后两人以 2.5 英里/小时的速度步行去饭店。他们提前了 5 分钟到达。 【问题】： Kellan 几点钟离开家？计算结果去掉时间表示形式中的冒号，如 5:30 表示成 530。 【解析】： 根据题意可得，本题需要分段求出相应的距离。 Step 1 根据第三题的计算结果可以发现 Andrea 到饭店的距离为 1.5 英里。因此两个人步行的时间为 $\frac{1.5}{2.5} = 0.6$ 小时 $= 36$ 分钟。 Step 2 Kellan 居住的地方离 Andrea 有 12 英里远，因此 Kellan 骑车的时间为 $\frac{12}{15} = 0.8$ 小时 $= 48$ 分钟。 Step 3 总计在路上花费的时间为 $36 + 48 = 84$ 分钟。他们 9 点开始吃早饭，由于提前 5 分钟到，在路上花费 84 分钟，所以倒推得到出发时间为 7:31，根据题目对最终数据的要求，得到 731。

5	45.5	【结构解析】: 【1】Bridget is starting a tutoring business to help adults get their GEDs. 【2】She already has five clients and decides they can share a single textbook, which will be kept at her office, and that she also needs one notebook and four pencils for each of them. 【3】She records her supply budget, which includes tax, in the table shown. 【4】The textbook makes up what percent of Bridget's total supply budget? 【5】Round to the nearest tenth of a percent and ignore the percent sign when entering your answer. 【1】：背景，可略读。【2-3】：条件。【4-5】：问题与要求。 【题干】： Bridget 开始一项辅导业务来帮助成人获得一般同等文凭。她已经有了五位顾客，并让他们共享一本教科书，这本书放在她的办公室保管。此外，她需要给每一位顾客配备一本笔记本和四支铅笔。她将供应预算列在下表中，其中包括了税价。 【问题】： 教科书的价格在她所列出的供应品清单中所占的比例为多少？（请保留到十分位，并且在填答案时忽略百分号） 【解析】： 根据题意可得，教科书的价格为 \$24.99，因此在整个供应品清单中所占的比例为：$\dfrac{24.99}{24.99\times1+3.78\times5+0.55\times20}\times100\%=45.527\%\approx45.5\%$，因此答案为 45.5。
6	100	【结构解析】: 【1】Bridget's business does very well, and she needs more supplies. 【2】She always orders them according to the table above, for five clients at a time. 【3】At the beginning of this year, she orders the supplies for the whole year, which cost \$988.02. 【4】Halfway through the year, she decides to take inventory of the supplies. 【5】She has used \$713.57 worth of the supplies. 【6】How many pencils should be left, assuming the supplies were used at the rate for which she originally planned? 【1-2】：背景，可略读，但是需要注意"以五个人为一个单位"。【3-5】：条件。【6】：问题。 【题干】： Bridget 的业务开展得很好，因此她需要扩大供应量。她总是按照表格中的信息来准备，一次对应五位顾客。在这一年年初的时候，她提供的本年度供应品预算为 \$988.02。半年后，她决定要清点一下供应品，发现已经花费了 \$713.57 在供应品上。 【问题】： 假设还是按照原始计划的使用率，则有多少支铅笔剩余？ 【解析】： 根据题意可得，按照原始计划的使用率，即教科书、笔记本和铅笔的使用比例保持为 1:5:20，这是完整的一组，后续的计算需要以此为单位进行整体计算。 Step 1 剩余的钱为：988.02 - 713.57 = 274.45。 Step 2 根据第五题的中间计算结果可得一套（含一本教科书、一本笔记本和二十支铅笔）的价格为 \$54.89。 Step 3 根据剩余的钱数可以得到剩余的套数为 $\dfrac{274.45}{54.89}=5$ 套，其中每套含铅笔 20 支，因此一共剩余铅笔 20×5 = 100 支。

| 7 | B | 【结构解析】:
【1】 The amount of glucose, or sugar, in a person's blood is the primary indicator of diabetes. 【2】 When a person fasts (doesn't eat) for eight hours prior to taking a blood sugar test, his/her glucose level should be below 100 mg/dL. 【3】 A person is considered at risk for diabetes, but is not diagnosed as diabetic, when fasting glucose levels are between 100 and 125. 【4】 If the level is above 125, the person is considered to have diabetes. 【5】 The table below shows the ages and glucose levels of a group of diabetes study participants. 【6】 According to the data, which age group had the smallest percentage of people with a healthy blood sugar level?
【1-2】：本题背景。【3-5】：条件。【6】：问题。
【题干】：
一个人血液中葡萄糖或糖的含量是表征糖尿病的主要指标。当一个人空腹（不吃食物）八小时之后再做血糖测试，他/她的血糖水平应低于 100 mg/dL。如果一个人的空腹血糖水平在 100～125 之间，则表明该测试者虽然不是糖尿病患者，但是有患糖尿病的风险。如果空腹血糖水平大于 125，这个人就可以被诊断为糖尿病患者。下面的表格表示一群糖尿病研究参与者年龄与血糖水平的关系。
【问题】：
根据表格的信息，哪一个年龄组中健康血糖水平所占的比例最小？
【解析】：
根据题意可得，健康血糖水平为 <100 mg/dL，因此可以从表中获得：
18～25 岁组：$\frac{9}{48} = 0.1875 = 18.75\%$；
26～35 岁组：$\frac{16}{98} = 0.1632 = 16.32\%$；
36～45 岁组：$\frac{19}{94} = 0.2021 = 20.21\%$；
大于 45 岁组：$\frac{12}{60} = 0.2 = 20\%$。
可见，其中健康血糖水平所占的比例最小的为 26～35 岁组，选 B。 |
| 8 | D | 【结构解析】:
【1】 Based on the table, if a single participant is selected at random from all the participants, 【2】 what is the probability that he or she will be at risk for diabetes and be at least 36 years old?
【1】：本题新增条件。【2】：问题。
【题干】：
根据表格中的信息，如果随机抽选一个研究参与者。
【问题】：
能抽到有患糖尿病风险，且年龄大于等于 36 岁的参与者相应的概率是多少？
【解析】：
根据题意可得，要同时满足两个条件：1) 有患糖尿病风险，即血糖水平为 100 mg/dL～125 mg/dL；2) 年龄≥36。从原题条件中可以得到满足这两个条件的人数为 35 + 27，因此抽到的概率为：
$\frac{35+27}{300} = \frac{31}{150}$，选 D。 |
| 9 | 2.25 | 【结构解析】:
【1】 Eli left his home in New York and traveled to Brazil on business. 【2】 On Monday, he used his credit card to purchase these pewter vases, R\$ 128, R\$ 66 and R\$ 85, as a gift for his wife. 【3】 For daily purchases totaling less than 200 U.S. dollars, Eli's credit card company charges a 2% fee. 【4】 If the total charge on his credit card for the vases was \$ 126.48, and no other purchases were made, 【5】 what was the foreign exchange rate on Monday in Brazilian reais (R\$) per U.S. dollar? 【6】 If necessary, round your answer to the nearest hundredth. |

第一部分　文字题：结构分析骨干提取法

9	2.25	【1】：条件，可略读。【2-4】：条件。【5-6】：问题与要求。 【题干】： Eli 离开他在纽约的家去巴西商务旅行。周一他用信用卡买了三个青灰色的花瓶作为礼物送给妻子，价格分别为 R\$ 128，R\$ 66 和 R\$ 85。如果每天的消费额不足 \$ 200，Eli 的信用卡公司需要收取 2% 的手续费。如果他使用信用卡为花瓶总计支付了 \$ 126.48，此外没有其他消费了。 【问题】： 周一，巴西雷亚尔对美元的外币兑换汇率为多少？（最终的数值保留到百分位） 【解析】： 根据题意可得，解答汇率问题的关键是找到同一类物品对应的不同货币类型以及货币量。 **Step 1** 三个青灰色花瓶的总价为：R\$ 128 + R\$ 66 + R\$ 85 = R\$ 279。 **Step 2** Eli 使用信用卡为花瓶总计支付了 \$ 126.48，由于没有超过 \$ 200，因此信用卡公司需要收取 2% 的手续费，所以 \$ 126.48 是含这一部分费用的。因此可以设三个花瓶的原价为 x，可得：$x \times (1 + 2\%) = \$ 126.48$，解得 $x = \$ 124$。说明买下这三个花瓶花费了 \$ 124，对应巴西货币雷亚尔为 R\$ 279。因此可得巴西雷亚尔对美元的外币兑换汇率为 $\frac{279}{124} = 2.25$。
10	3	【结构解析】： 【1】On Wednesday, Eli bought a tourmaline ring that cost R\$ 763. 【2】For daily purchases over \$ 200, his credit card company charges the same 2% fee on the first \$ 200 of the converted price and 3% on the portion of the converted price that is over \$ 200. 【3】If the total charge on his credit card for the ring was \$ 358.50, 【4】what was the amount of decrease, as a percentage, in the foreign exchange rate between Monday and Wednesday? 【5】Round your answer to the nearest whole percent. 【1-3】：条件。【4-5】：问题与要求。 【题干】： 周三的时候，Eli 购买了一枚电气石钻戒，价格为 R\$ 763。对于日消费额超过 \$ 200，他的信用卡公司会对 \$ 200 部分征收 2% 的手续费，对超过 \$ 200 的部分征收 3% 的手续费。如果他购买电气石钻戒总计用信用卡花费额度为 \$ 358.50。 【问题】： 从周一到周三外汇汇率下降的百分比为多少？（结果保留至整数） 【解析】： 根据题意可得，对于日消费额超过 \$ 200，他的信用卡公司会对 \$ 200 部分征收 2% 的手续费，而对超过 \$ 200 的部分征收 3% 的手续费。说明需要对费用进行分段计算。 **Step 1** 设他购买的电气石钻戒价格为 \$ x，可得：$200 \times (1 + 2\%) + (x - 200) \times (1 + 3\%) = 358.50$，计算得到 $x = \$ 350$。 **Step 2** 电气石钻戒的价格为 \$ 350，即 R\$ 763，因此可得周三外汇汇率为 $\frac{763}{350} = 2.18$。 **Step 3** 从第九题可知周一的外汇汇率为 2.25，因此从周一到周三外汇汇率下降的百分比为 $\frac{2.25 - 2.18}{2.25} \times 100\% = 3.11\% \approx 3\%$。

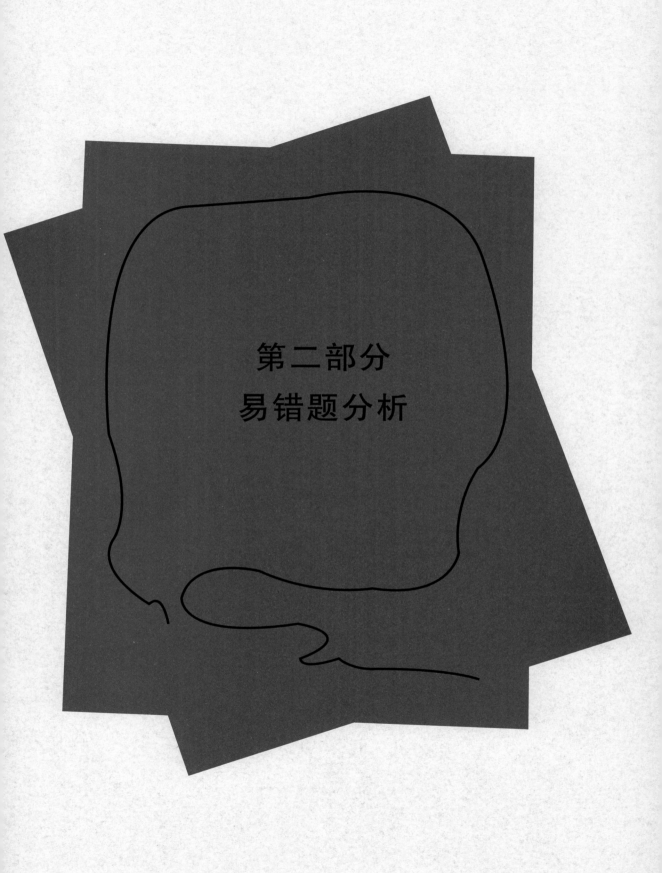

第二部分
易错题分析

第二部分　易错题分析

1. For the equation $\sqrt{mx-5} = x+3$, the value of m is -3. What is the solution set for the equation?
   A. $(-3, 3)$　　　B. (-2)　　　C. $(-2, -7)$　　　D. $(3, 6)$

答案	B
知识点	考查解方程的基本过程。
解析	【题干】：如果满足方程式 $\sqrt{mx-5} = x+3$，并且 m 的值为 -3。 【问题】：该方程的解为多少？ 【解析】：Step 1：将 $m=-3$ 代入。Step 2：方程两边都平方得到 $-3x-5 = x^2+6x+9$。Step 3：解得 $x_1 = -2$，$x_2 = -7$。 注意原题中 $x+3$ 必须大于 0，因此需要舍去 $x=-7$，选 B。
易错点	此题易错选为 C。学生在解题时首先会得到两个解，就容易被选项引导。注意解此类题时要将所有的解代入确认是否满足原题的要求。
拓展	类似考点：方程的解代入原式满足分母不为 0。
必背词汇	equation：n. 方程式；等式 solution set：解集，解集合

2. Jan's rectangular baking sheet is 9.5 inches (in) by 13 in. To the nearest inch, what is the longest bread stick Jan could bake on his baking sheet?
   A. 13 in　　　　B. 16 in　　　　C. 124 in　　　　D. 259 in

答案	B
知识点	考查勾股定理在实际生活中的运用。
解析	【题干】：Jan 的矩形烤盘尺寸为 9.5 英寸×13 英寸。 【问题】：Jan 可以烤的长棍面包最长为多少？ 【解析】：baking sheet 指的是烤盘，尺寸为 9.5 英寸×13 英寸。题目问的是可以放多长的 bread stick（长棍面包）。所谓"最长"应该是指倾斜放可以取得的最大值。这道题转化为数学语言就是在一个直角三角形中，如果知道了两个直角边分别是 9.5 和 13，求斜边的长度。根据勾股定理得到斜边长为 16.10≈16，选 B。
易错点	此题易错选为 A，学生在快速读题时容易直接想到最长的边就是 13 英寸，所以就认为最长可以放 13 英寸的长棍面包。
拓展	学生在读题过程中一定要判断该题的考核要求是什么，新 SAT 数学部分的两大考核要求是： 1) 考查考生对基本概念的理解； 2) 借助工具迅速地实现对数据的分析和处理。 此题的解答虽然需要用到计算器，但主要还是考查了勾股定理在实际生活中的运用。如果给出的两个条件是 5 和 12 英寸，那么学生可能很快就会联想到直角三角形。
必背词汇	rectangular：adj. 矩形的；成直角的

3. Lily gives his little sister Lucy a 15 second (sec) head start in their 300 meter (m) race. During the race, Lucy runs at an average speed of 5 m/sec and Lily runs at an

average speed of 8 m/sec. Which of the following best approximates the number of seconds that Lily will run before he catches Lucy?

A. 5
B. 25
C. 40
D. 55

答案	B
知识点	考查方程在实际生活中的运用。
解析	【题干】：在300米比赛中，Lily让她的小妹妹Lucy先跑15秒。在整个比赛中，Lucy跑的平均速度为5米/秒，Lily跑的平均速度为8米/秒。 【问题】：下面哪个（最接近的秒数）是Lily追上Lucy用的时间？ 【解析】：head start指领先，说明Lucy先跑15秒，Lucy的速度是5米/秒，Lily的速度是8米/秒。设所需要的时间为t，则满足这样的方程关系：$8t = 5(t+15)$，解得$t = 25$，选B。
易错点	此题学生在读题时容易先入为主地从300米这个条件开始，从而导致无从下手。
拓展	运动问题包含的条件一般有速度和时间，而不同人或者物的关系是速度×时间即总路程相等。
必背词汇	head start：领先 approximate：adj. 近似的；大概 vt. 近似；使……接近；粗略估计 vi. 接近于；近似于

4.
$$x - y = 3$$
$$2x = 2y + 6$$

The system of equations above has how many solutions?

A. exactly one
B. exactly two
C. exactly three
D. infinitely many

答案	D
知识点	考查代数关系。
解析	将式2进行转化后可得$2x - 2y = 6$，化简为$x - y = 3$，这样发现两个式子其实为一个式子。说明这个方程组有无数解，选D。
易错点	此题学生会认为二元一次方程组必然有一组对应的解，就直接选A，忽视了两个方程组其实就是一个式子。
必背词汇	equation：n. 方程式；等式 solution：n. 解答；解决方案；溶液；溶解

5. The cost C of manufacturing a certain product can be estimated by the formula $C = 0.03rst^2$, where r and s are the amounts, in pounds, of the two major ingredients and t is the production time in hours. If r is increased by 50 percent, s is increased by 20 percent, and t is decreased by 30 percent, by approximately what percent will the estimated cost of manufacturing the product change?

A. 40% decrease
B. 12% increase
C. 4% increase
D. 12% decrease

答案	D
知识点	考查代数中的百分比关系。
解析	【题干】：产品的成本 C 可以用 $C=0.03rst^2$ 表示，其中 r 和 s 是该产品两种主要成分的磅数，t 是产品生产的小时数，如果 r 增加 50%，s 增加 20%，而 t 减少 30%。 【问题】：产品的生产成本将改变多少？（以百分比计） 【解析】：根据题意发现变化后的 r、s 和 t 分别为 $1.5r$、$1.2s$ 和 $0.7t$，则产品的成本变化百分比为： $\dfrac{0.03rst^2 - 0.03 \times 1.5r \times 1.2s \times (0.7t)^2}{0.03rst^2} \times 100\%$ $= \dfrac{0.03rst^2(1 - 1.5 \times 1.2 \times 0.7^2)}{0.03rst^2} \times 100\% = 11.8\% \approx 12\%$ 由于 $0.03 \times 1.5 \times 1.2 \times 0.7^2 < 1$，所以是减少了，选 D。
易错点	此题学生会认为 r 和 s 都是增加，所以总的成本也相应增加。
必背词汇	estimate：n. 估计；估价；判断；看法；vt. 估计；估量；判断；评价 formula：n. 公式 amount：n. 数量；总额；总数；vi. 总计；合计；相当于；共计；产生……结果 ingredient：n. 原料；要素；组成部分；adj. 构成组成部分的

6. For a group of n people, k of whom are of the same sex, the expression $\dfrac{n-k}{n}$ yields an index for a certain phenomenon in group dynamics for members of that sex. For a group that consists of 20 people, 4 of whom are females, by how much does the index for the females exceed the index for the males in the group?

答案	0.6
知识点	考查文字表述中的代数关系。
解析	【题干】：在一群由 n 个人组成的人群中，有 k 个人是同性别的，表达式 $\dfrac{n-k}{n}$ 表示的是该性别成员在群体中的动态变化（即性别指数），如果在一组人群中有 20 个人，其中有 4 个女性。 【问题】：该群人中女性指数超出男性指数多少？ 【解析】：根据题意可得，20 人中女性有 4 人，男性有 16 人，则： 女性指数为：$\dfrac{n-k}{n} = \dfrac{20-4}{20} = 0.8$，而男性指数为：$\dfrac{n-k}{n} = \dfrac{20-16}{20} = 0.2$，所以差异为 $0.8 - 0.2 = 0.6$。
易错点	此题学生会算出女性指数，但会忽略问题需要计算出超过的部分。
必背词汇	expression：n. 表达式；表达 yield：n. 产量；收益；vt. 屈服；出产；产生；放弃；vi. 屈服；投降 index：n. 指标；指数；索引；指针；vt. 指出；编入索引中；vi. 做索引 phenomenon：n. 现象 dynamics：n. 动力学 exceed：vt. 超过；胜过

7. Mr. Kramer, the losing candidate in a two-candidate election, received 942,568 votes, which was exactly 40 percent of all the votes cast. Approximately what percent of the remaining votes would he need to have received in order to have won at least 50 percent of all the votes cast?

A. 10%
B. 12%
C. 15%
D. 17%

答案	D
知识点	考查文字表述中的代数关系。
解析	【题干】：Karmer 是两个候选人中的落选者，获得了 942,568 张选票，恰好占总投票数的 40%。 【问题】：为了能至少获得总选票的 50%，他还需要从其余的选票中得到大约百分之多少的选票？ 【解析】：根据题意，设还需要获得选票的百分比为 x，因此可得 $40\% + (1-40\%) \times x \geq 50\%$，解得 $x \geq 16.7\%$，选 D。
易错点	此题学生会纠结在 942,568 这个值，其实在本题中这个条件根本不需要。
必背词汇	exactly：adv. 恰好地；正是；精确地；正确地 approximately：adv. 大约；近似地；近于

8. The present value (PV) of an investment is the amount that should be invested today at a specified interest rate in order to earn a certain amount at a future date. The amount desired is called the future value. Approximately how much should be invested today in a savings account that earns 3% interest compounded annually in order to have $500 in 2 years?

A. $515
B. $470
C. $485
D. $530

答案	B
知识点	考查文字表述中的指数关系。
解析	【题干】：投资现值是指今天投入的资金量。这个投资可以按照一定的利率在未来某天获得相应的金额，这个数量称为未来值。 【问题】：请确定今天的投资量大约是多少，使得如果当采用的利率是 3% 的年复利，2 年后可以获得 $500？ 【解析】：根据题意设今天的投资量为 x，可得： $x \times (1+3\%) \times (1+3\%) = 500$，解得 $x = 471.30 \approx 470$，选 B。
易错点	在此题中，学生没有理解什么是复利率计算。设今天的投资为 x，则第一年可以获得的金额为 $x \times (1+3\%)$，第二年是在此基础上再乘以相应的利率，因此 $x \times (1+3\%) \times (1+3\%)$。
必背词汇	investment：n. 投资；投入 specified interest rate：指定利率；固定利率 compounded annually：按年复利计算；每年累计计算

9. Which quadrants contain the solutions to this system of inequalities?

$$y - 2x \leqslant -3$$
$$3y - x \geqslant -4$$

A. quadrants Ⅰ and Ⅳ B. quadrants Ⅱ and Ⅲ
C. quadrants Ⅲ and Ⅳ D. quadrants Ⅱ，Ⅲ and Ⅳ

答案	A
知识点	考查不等式的概念。
解析	【题干】：给出了一个不等式组。 【问题】：不等式组的解在坐标图中第几象限？ 【解析】：如图 2-1，根据不等式的定义可得，如果 $y > kx + b$，则 y 的取值范围在直线 $y > kx + b$ 图像的上方，类似地，$y < kx + b$，则在直线图像的下方。 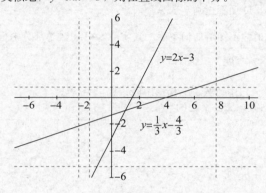 图 2-1 根据图可知，需要找到 $y = 2x - 3$ 的图像下方和 $y = \frac{1}{3}x - \frac{4}{3}$ 图像上方相交的位置，即处在第 Ⅰ、Ⅳ 象限，选 A。
易错点	不等式的问题让很多学生感到困难，其实只要将不等式转化为坐标图中的直线图像，再根据不等式符号观察是在图像的上方还是下方即可判断。
必背词汇	quadrant：n. 象限

10. A set of data has 10 values，no two of which are the same. If the smallest value is removed from the set，which of the following statements MUST be true?
A. The range of the first data set is greater than the range of the second data set
B. The mode of the first data set is greater than the mode of the second data set
C. The medians of the two data sets are the same
D. The mean of the first data set is greater than the mean of the second data set

答案	A
知识点	考查统计学中的平均值、中位数、众数、取值范围等概念。
解析	【题干】：一个数列中有10个数，每一个数都不相同。 【问题】：如果把该数列中最小的数移除之后，下列哪一项表述是正确的？ A. 之前数列的数值范围比之后数列的大 B. 之前数列的众数比之后数列的大 C. 两个数列的中位数相等 D. 之前数列的平均数比之后数列的大 【解析】：根据题意可以举例判别。如假设该数列为1，2，3，4，5，6，7，8，9，10。首先看一下数值范围变化，之前数列为10−1=9；之后数列为10−2=8，之前数列的数值范围比之后数列的大，A项正确。由于原数列中没有一个重复的数值，所以原数列和之后的数列不存在众数，B项不正确。中位数：之前数列的中位数为5.5；之后数列的为6，两者不相等，C项不正确。平均值：之前数列的平均值为5.5；之后数列的为6，之前数列的平均数比之后数列的小，D项不正确，选A。
易错点	很多同学对这道题不知道如何下手，其实新SAT-1数学部分中很多题可以采用举例代入法。
必背词汇	range：n. 范围；幅度 mode：n. 模式；众数 median：n. 中值；中位数；三角形中线；梯形中位线 mean：n. 平均值；adj. 平均的

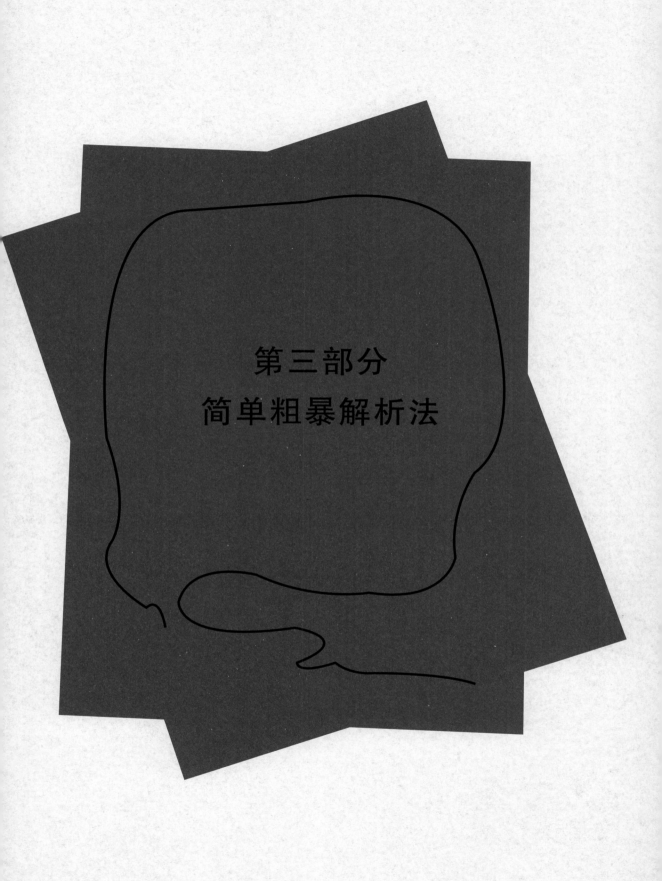

第三部分
简单粗暴解析法

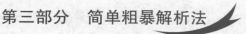

1. If the expression $\dfrac{9x^2}{3x-2}$ is written in the equivalent form $\dfrac{4}{3x-2}+k$, what is k in terms of x?

A. $12x-8$
B. $4x^2-4$
C. $9x^2$
D. $3x+2$

答案	D
解析	【题干】：如果方程式 $\dfrac{9x^2}{3x-2}$ 可以写成 $\dfrac{4}{3x-2}+k$ 的形式。 【问题】：k 用 x 来表示是下面哪一个式子?
巧解	在计算器①的菜单中选择：[3-代数] → [1-求解] 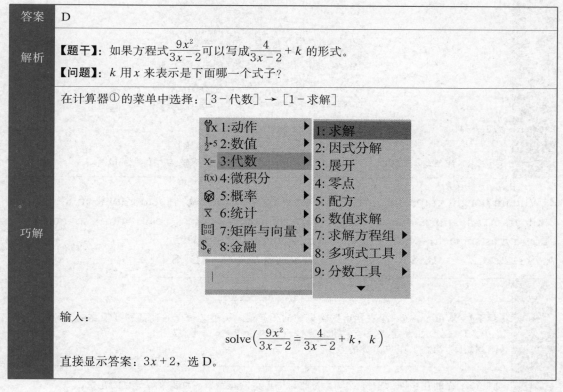 输入： $$\text{solve}\left(\dfrac{9x^2}{3x-2}=\dfrac{4}{3x-2}+k,\ k\right)$$ 直接显示答案： $3x+2$，选 D。

2.
$$f(x)=-2x^2-k$$

The equation of a quadratic function lying in the xy-plane is given above, where k is a constant. If （-5，-32）is a point on the curve, what is the maximum value of $f(x)$?

A. 16
B. -16
C. -18
D. 18

答案	D
解析	【题干】：如果该方程在 xy 坐标系中，k 为一个定值，且（-5，-32）是图像上的一个点。 【问题】：$f(x)$ 的最大值为多少？

---

① 本部分计算器型号为 TI-Nspire CX-CCAS，为 CollegeBoard 所认可，详细信息参见官方网址：https://collegereadiness.collegeboard.org/sat/taking-the-test/calculator-policy

巧解	在计算器的菜单中选择：[3-代数] → [1-求解] 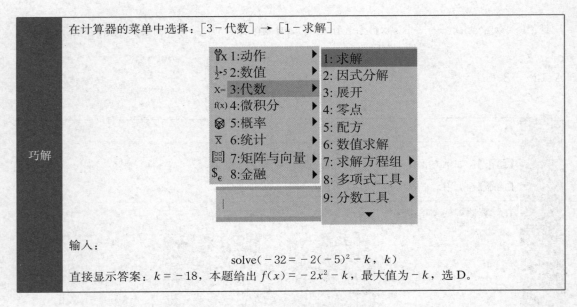 输入： $$\text{solve}(-32 = -2(-5)^2 - k, k)$$ 直接显示答案：$k = -18$，本题给出 $f(x) = -2x^2 - k$，最大值为 $-k$，选 D。

3. William bought grapes and spinach at a farmer's market. He spent a total of \$13.20 on 2 kilograms of grapes and 5 kilograms of spinach. If grapes cost 3 times as much per kilogram as spinach, what is the cost per kilogram of grapes?

A. \$1.20 B. \$2.40 C. \$3.60 D. \$4.80

答案	C
解析	【题干】：William 在一家农贸市场购买葡萄和菠菜。买了 2 千克的葡萄和 5 千克的菠菜，一共花费 \$13.20。如果葡萄每千克的花费是菠菜的 3 倍。 【问题】：葡萄每千克的花费是多少？
巧解	葡萄每千克的花费即为单价。设 g 为葡萄每千克的花费，s 为菠菜每千克的花费。 在计算器中选择：[3-代数] → [7-求解方程组] 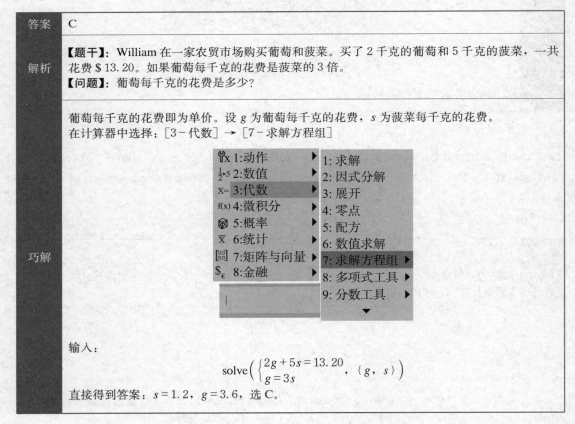 输入： $$\text{solve}\left(\begin{cases} 2g + 5s = 13.20 \\ g = 3s \end{cases}, \{g, s\}\right)$$ 直接得到答案：$s = 1.2$，$g = 3.6$，选 C。

4. In a math competition, a student can solve algebra or geometry problems. The student earns 25 points each algebra problem solved and 40 points for each geometry problem solved. Alessia solved a total of 50 problems, scoring 1,475 points in all. How many geometry problems did Alessia answer correctly?

A. 10   B. 15   C. 25   D. 35

答案	B
解析	【题干】：在一次数学竞赛中，学生需要解答代数和几何两种题目。学生每解答一道代数题可以得 25 分，解答一道几何题可以得 40 分。Alessia 一共解答出了 50 道题，获得了 1,475 分。 【问题】：Alessia 解答了多少道几何题？
巧解	设解决的代数题数量为 a，解决的几何题数量为 g 在计算器中选择：[3-代数] → [7-求解方程组] 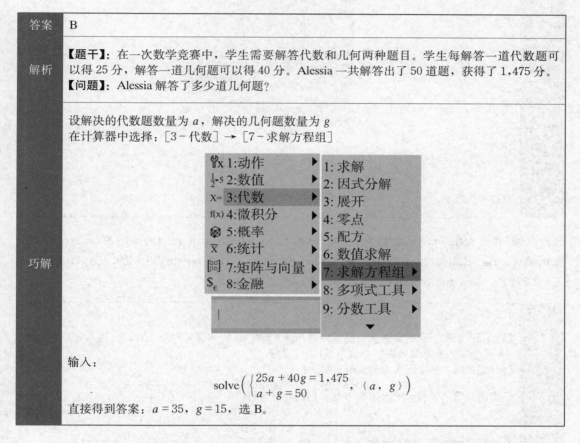 输入：$\text{solve}\left(\begin{cases} 25a + 40g = 1{,}475 \\ a + g = 50 \end{cases}, \{a, g\}\right)$ 直接得到答案：$a = 35$，$g = 15$，选 B。

5. An art museum offers discounted tickets to seniors, and children do not need a ticket to enter. A group visiting the museum included 64 adults, 44 children, and 16 seniors. A second group included 48 adults, 56 children, and 16 seniors. If the museum collected $548 in ticket fees from the first group and $428 from the second group, what was the price, in dollars, of an adult ticket?

答案	7.5
解析	【题干】：一家艺术博物馆为老年人提供打折门票，而儿童可以免票进入。一个旅行团有 64 位成人，44 位孩子和 16 位老人。另一个旅行团有 48 位成人，56 位孩子和 16 位老人。该博物馆向第一家旅行团收取门票费为 $548，向第二家旅行团收取门票费为 $428。 【问题】：一个成人的票价为多少？

巧解	设成人票价为 a，老人票价为 s，根据题意可得： $$64a + 16s = 548$$ $$48a + 16s = 428$$ 在计算器中选择：［3－代数］→［7－求解方程组］ 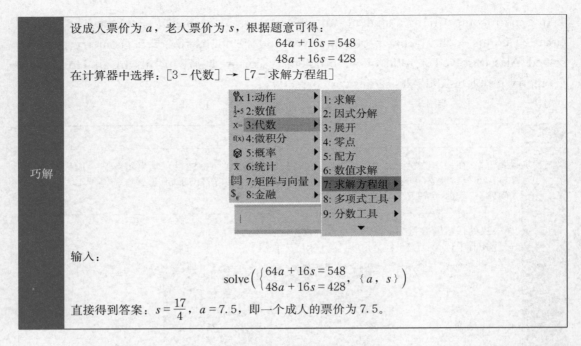 输入： $$\text{solve}\left(\begin{cases}64a + 16s = 548 \\ 48a + 16s = 428\end{cases}, \{a, s\}\right)$$ 直接得到答案：$s = \dfrac{17}{4}$，$a = 7.5$，即一个成人的票价为 7.5。

6. A local deli sells sandwiches for ＄4.50 each and bottles of ice tea for ＄2.25 each. On Saturday, the deli sold a total of 307 sandwiches and bottles of ice tea, taking in ＄1,077.75. How many sandwiches were sold on Saturday?

答案	172
解析	【题干】：当地一家熟食店卖三明治，每份＄4.50，瓶装冰茶，每瓶＄2.25。周六该店共卖出 307 份三明治和瓶装冰茶，总价为＄1,077.75。 【问题】：周六该家店卖出多少份三明治？
巧解	设卖出三明治 s 份，卖出冰茶 b 瓶，根据题意可得： $$4.50s + 2.25b = 1,077.75$$ $$s + b = 307$$ 在计算器中选择：［3－代数］→［7－求解方程组］ 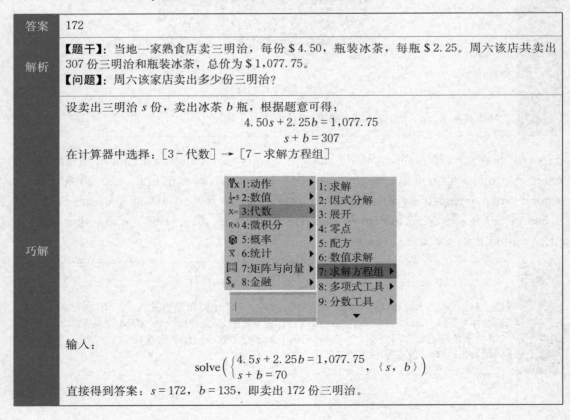 输入： $$\text{solve}\left(\begin{cases}4.5s + 2.25b = 1,077.75 \\ s + b = 70\end{cases}, \{s, b\}\right)$$ 直接得到答案：$s = 172$，$b = 135$，即卖出 172 份三明治。

7. A ceramics factory ships teacups in cartons that weigh 23 pounds each, and dessert plates in cartons that weigh 45 pounds each. A shipment going to several department stores weighs 2,314 pounds in total and contains 70 cartons. How many cartons of teacups are in the shipment?

答案	38
解析	【题干】：一家陶瓷厂需要将茶杯装箱船运，每箱重达23磅。将甜点盘装箱船运，每箱重达45磅。有批船运的货送到不同的商店，总重量为2,314磅，共计装了70箱。 【问题】：茶杯一共装了几箱？
巧解	设茶杯一共装了 t 箱，甜点盘装了 d 箱，根据题意可得： $$23t + 45d = 2,314$$ $$t + d = 70$$ 在计算器中选择：[3-代数] → [7-求解方程组] 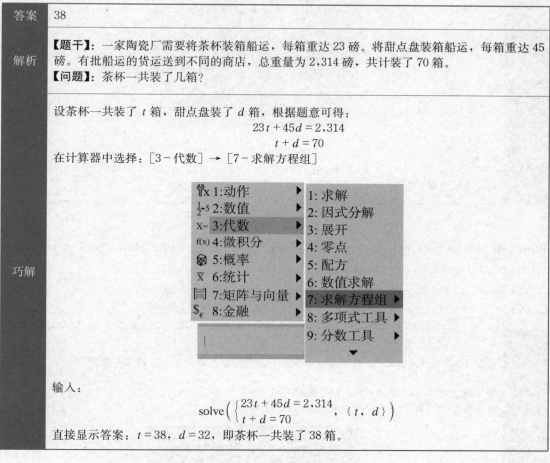 输入：$\mathrm{solve}\left(\begin{cases}23t + 45d = 2,314 \\ t + d = 70\end{cases}, \{t, d\}\right)$ 直接显示答案：$t = 38$，$d = 32$，即茶杯一共装了38箱。

8. $(3w-5)^2 - (3w-5)$

Which of the following is an equivalent expression?

A. $3w - 5$
B. $3w^2 - 27w + 20$
C. $9w^2 - 33w - 30$
D. $3(3w-5)(w-2)$

答案	D
解析	$(3w-5)^2 - (3w-5)$ 可以写成下面哪一个式子？

巧解	在计算器中选择：[3-代数] → [2-因式分解] 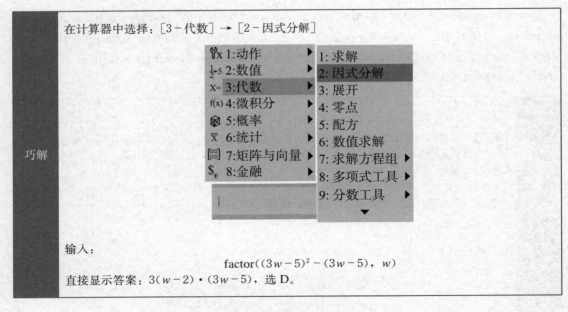 输入： $$factor((3w-5)^2-(3w-5), w)$$ 直接显示答案：$3(w-2) \cdot (3w-5)$，选 D。

9. $$f(x) = x^2 + 4(x-3)$$
Which of the following is an equivalent form of the expression above that displays the zeroes of the function as constants in the expression?
A. $f(x) = x^2 - 4x - 12$
B. $f(x) = (x-6)(x+2)$
C. $f(x) = (x+6)(x-2)$
D. $f(x) = (x+2)^2 - 16$

答案	C
解析	已知 $f(x) = x^2 + 4(x-3)$，下面哪一个式子表示的是在该式子中是以零为常数？
巧解	根据题意可得，以零为常数，就是需要写成 $(x-a) \times (x-b)$ 的形式 在计算器中选择：[3-代数] → [2-因式分解] 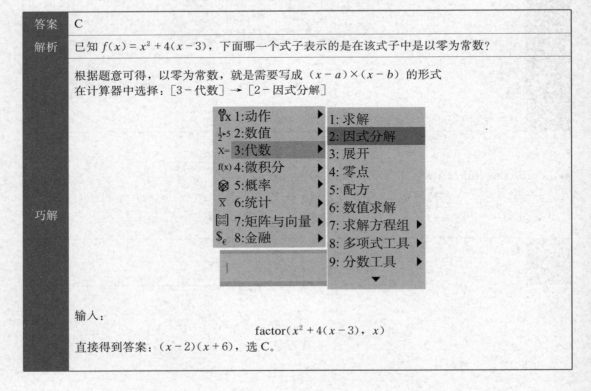 输入： $$factor(x^2+4(x-3), x)$$ 直接得到答案：$(x-2)(x+6)$，选 C。

10. $$f(x) = (x+10)(x-6)$$
Which of the following is an equivalent form of the function f above in which the minimum value of f appears as a constant?
A. $x^2 - 60$
B. $x^2 + 40x - 60$
C. $(x-2)^2 - 64$
D. $(x+2)^2 - 64$

答案	D
解析	已知 $f(x) = (x+10)(x-6)$，下面哪一个式子表示的是函数 f 的最小值为一个定值？
巧解	根据题意可得，函数的最小值为一个定值，即写成 $a(x-b)^2 + c$ 的形式，c 是函数的最小值，且是一个常数。 在计算器中选择：[3-代数] → [5-配方] 1:动作　　　　1:求解 2:数值　　　　2:因式分解 3:代数　　　　3:展开 4:微积分　　　4:零点 5:概率　　　　5:配方 6:统计　　　　6:数值求解 7:矩阵与向量　7:求解方程组 8:金融　　　　8:多项式工具 　　　　　　　9:分数工具 输入： complete square$((x+10)(x-6), x)$ 直接得到结果：$(x+2)^2 - 64$，即最小值恒定为 -64，选 D。

11. $$y = 2x^2 - 2x - 12$$
Which of the following is an equivalent form of the above equation of the parabola in the xy-plane, from which the coordinates of its vertex can be identified as constants in the equation?

A. $y = (2x-4)(x+3)$
B. $y = 2\left(x - \dfrac{1}{2}\right)^2 - \dfrac{25}{2}$
C. $y = 2x(x-2) - \dfrac{25}{2}$
D. $y = 2\left(x + \dfrac{1}{2}\right)^2 - \dfrac{25}{2}$

答案	B
解析	已知 $y = 2x^2 - 2x - 12$，下面哪一个式子在平面直角坐标系中能够表现出顶点是一个定值的形式？

根据题意可得，函数的顶点为一个定值，即写成 $a(x-b)^2+c$ 的形式，顶点坐标为 (b, c)。
在计算器中选择：[3-代数] → [5-配方]

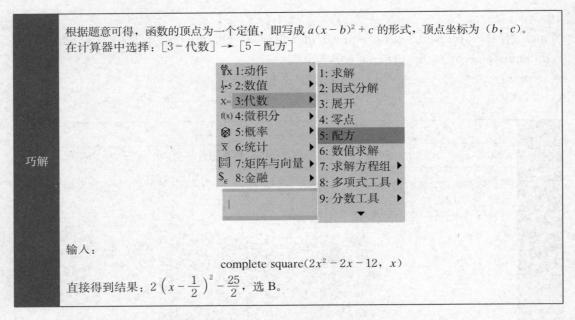

输入：
$$\text{complete square}(2x^2-2x-12, x)$$

直接得到结果：$2\left(x-\dfrac{1}{2}\right)^2-\dfrac{25}{2}$，选 B。

12. Joan sells hand-painted ceramic vases in her pottery shop. If the price of a vase is $20, Joan sells 4 vases per month. For every $1 decrease in the price of a vase, Joan sells 2 additional vases per month. To maximize the amount of a money Joan takes in per month from the sale of vases, how much should she charge for 1 vase?

答案	11
解析	【题干】：Joan 在她的陶瓷店售卖手绘陶瓷花瓶。如果每一个陶瓷花瓶价格为 $20，每月能卖出 4 只，如果价格每下降 $1，则每月可以多卖出 2 个。 【问题】：要想尽可能卖出更多的陶瓷花瓶，Joan 需要如何来定价？

巧解

设 n 为 Joan 降价的次数（每次降价 $1）。花瓶的原始价格为 20，因此降价 n 次后的价格为 $20-n$，花瓶售出的数量为 $4+2n$。
在计算器中选择：[3-代数] → [5-配方]

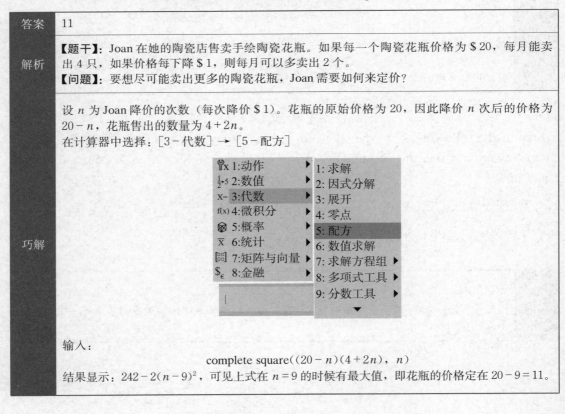

输入：
$$\text{complete square}((20-n)(4+2n), n)$$

结果显示：$242-2(n-9)^2$，可见上式在 $n=9$ 的时候有最大值，即花瓶的价格定在 $20-9=11$。

13. Philip sells Black Forest cakes in his bakery shop. If the price of a Black Forest cake is $25, Philip sells 10 Black Forest cakes per week. For every $2.50 decrease in the price of a Black Forest cake, Philip sells 5 additional Black Forest cakes per week. To maximize the amount of a money Philip takes in per week from the sale of Black Forest cakes, how much should he charge for 1 cake?

答案	15
解析	**【题干】**：Philip 的面包房售卖黑森林蛋糕，每份售价为 $25，每周可以卖出 10 份。如过每份价格每下降 $2.50，则每周可以多卖出 5 份。 **【问题】**：Philip 该如何定价使得每周的收益最大？
巧解	设 n 为 Philip 降价的次数（每次降价 $2.50）。蛋糕的原始价格为 25，因此降价 n 次后的价格为 $25-2.50n$，蛋糕售出的数量为 $(10+5n)$。 在计算器中选择：[3-代数] → [5-配方] 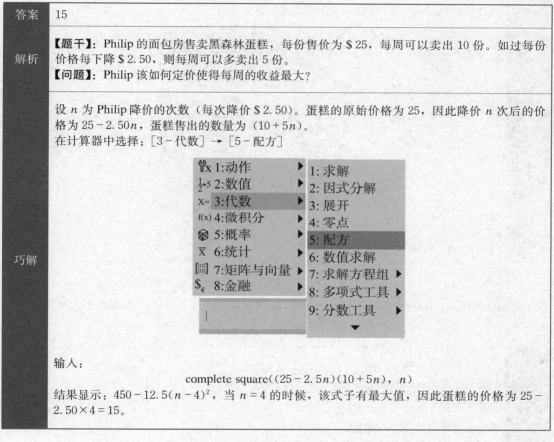 输入： $$\text{complete square}((25-2.5n)(10+5n), n)$$ 结果显示：$450-12.5(n-4)^2$，当 $n=4$ 的时候，该式子有最大值，因此蛋糕的价格为 $25-2.50\times 4=15$。

14. $$h = -16t^2 + v_0 t + s_0$$

The equation above expresses the approximate height h, in feet, of a ball t seconds after it is launched vertically upward from an initial height of s_0 feet with an initial velocity of v_0 feet per second. If the ball is launched from a height of 6 feet with a velocity of 40 feet per second, which of the following represents the number of seconds from launch it takes the ball to hit the ground, to the nearest tenth of a second?

A. 2.0
B. 2.5
C. 2.6
D. 3.0

答案	C
解析	【题干】：上式描述的是一个球在离地面 s_0 英尺的高度以初速度为 v_0 英尺/秒竖直上抛 t 秒后的高度。如果该球在离地面 6 英尺的高度以初速度为 40 英尺/秒竖直上抛。 【问题】：下面哪一个式子表示的是该球落到地面的时间？（数值保留到十分位）
巧解	问球落到地面的时间，即求当 $h=0$ 的时候 t 的数值。 在计算器中选择：[3-代数] → [2-因式分解] 1:动作　　1:求解 2:数值　　2:因式分解 3:代数　　3:展开 4:微积分　4:零点 5:概率　　5:配方 6:统计　　6:数值求解 7:矩阵与向量　7:求解方程组 8:金融　　8:多项式工具 　　　　　9:分数工具 输入： $$\text{solve}(-16t^2+40t+6=0,\ t)$$ 得到：$t=-0.141\,941$ or $t=2.641\,94$，结果取正值，$t\approx 2.6$，选 C。

15. $$x^2+10x+y^2-14y=-38$$

The equation of a circle in the xy-plane is shown above. What is the radius of the circle?

A. 5　　　B. 6　　　C. 25　　　D. 36

答案	B
解析	【题干】：上式表示的是在平面直角坐标系中圆的方程。 【问题】：该圆的半径为多少？
巧解	在计算器中选择：[3-代数] → [5-配方] 1:动作　　1:求解 2:数值　　2:因式分解 3:代数　　3:展开 4:微积分　4:零点 5:概率　　5:配方 6:统计　　6:数值求解 7:矩阵与向量　7:求解方程组 8:金融　　8:多项式工具 　　　　　9:分数工具 输入： $$\text{complete square}(x^2+10x+y^2-14y=-38,\ x,\ y)$$ 得到：$(x+5)^2+(y-7)^2=36$，根据圆的标准方程的定义得到 $36=r^2$，圆的半径 r 为 6，选 B。

16. $$2x^2 + 2y^2 + 12x - 24y + 60 = 0$$

The equation of a circle in the xy-plane is shown above. Which of the following statements is true?

A. The coordinates of the center are $(3, -6)$ and the length of the radius is 30

B. The coordinates of the center are $(-3, 6)$ and the length of the radius is $\sqrt{30}$

C. The coordinates of the center are $(3, -6)$ and the length of the radius is 15

D. The coordinates of the center are $(-3, 6)$ and the length of the radius is $\sqrt{15}$

答案	D
解析	【题干】：在平面直角坐标系中圆的标准方程如上式所示。 【问题】：下面哪一个表述是正确的？
巧解	在计算器中选择：［3－代数］→［5－配方］ 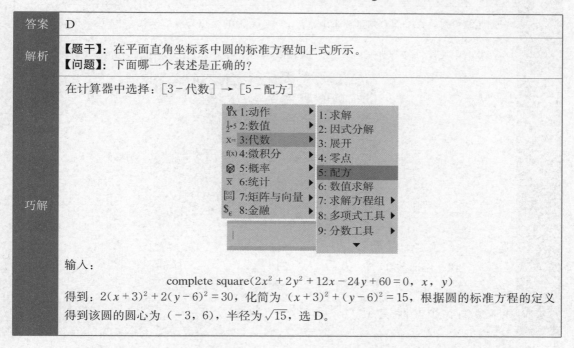 输入： 　　complete square$(2x^2 + 2y^2 + 12x - 24y + 60 = 0, x, y)$ 得到：$2(x+3)^2 + 2(y-6)^2 = 30$，化简为 $(x+3)^2 + (y-6)^2 = 15$，根据圆的标准方程的定义得到该圆的圆心为 $(-3, 6)$，半径为 $\sqrt{15}$，选 D。

17. If $f(x) = 4x - 5$ and $g(x) = 5x^2$, then what is the value of $f(g(2))$?

答案	75
巧解	在计算器中选择：［1－动作］→［1－Define］ 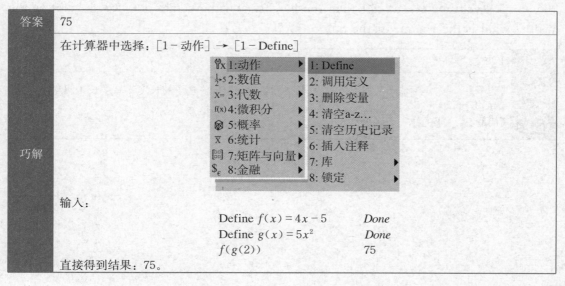 输入： 　　Define $f(x) = 4x - 5$　　Done 　　Define $g(x) = 5x^2$　　Done 　　$f(g(2))$　　　　　　　75 直接得到结果：75。

18.
$$f(x) = (x+3)^2 + k$$

The function $f(x)$ defined above lies in the xy-plane. If $f(x)$ crosses the x-axis at $(-8, 0)$ and at $(s, 0)$, what is the value of s?

答案	2
解析	【题干】：函数 $f(x)$ 如上式所示，在平面直角坐标系中，$f(x)$ 与 x 轴的交点为 $(-8, 0)$ 和 $(s, 0)$。 【问题】：s 的值为多少？
巧解	根据题意可得，在平面直角坐标系中，$f(x)$ 与 x 轴的交点为 $(-8, 0)$ 和 $(s, 0)$，即表明 $x = -8$ 和 $x = s$。 在计算器中选择：[1-动作] → [1-Define] 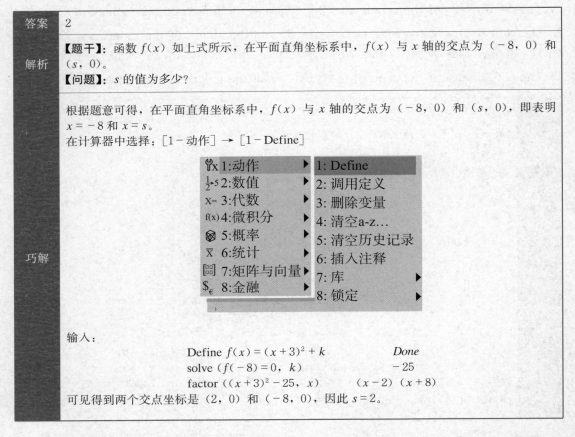 输入： Define $f(x) = (x+3)^2 + k$ Done solve $(f(-8) = 0, k)$ -25 factor $((x+3)^2 - 25, x)$ $(x-2)(x+8)$ 可见得到两个交点坐标是 $(2, 0)$ 和 $(-8, 0)$，因此 $s = 2$。

19. The bottom of the rectangular box is 96 square centimeters in area. If the length of the box is 4 centimeters greater than the height, and the height is twice as great as the width, what is the volume of the box?

答案	1,152
解析	【题干】：一个矩形盒子的底部面积为 96 平方厘米，如果该盒子的长度比高度长 4 厘米，高是宽的 2 倍。 【问题】：盒子的体积是多少？

| 巧解 | 设盒子的高度为 h，因此长度 $l = h + 4$，宽度 $w = \frac{1}{2}h$。
在计算器中选择：[3-代数] → [7-求解方程组]
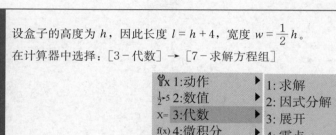
输入：$$\text{solve}\begin{cases}(h+4)\cdot\dfrac{h}{2}=96\\ v=(h+4)\cdot\left(\dfrac{h}{2}\right)\cdot h\end{cases}\{h,v\}$$
直接得到结果为：$v = -1,536$，$h = -16$ 或者 $v = 1,152$，$h = 12$，取 $v = 1,152$。|

20. A deposit of ＄800 is made into an account that earns 2% interest compounded annually. If no additional deposits are made, how many years will it take until there is ＄990 in the account?

A. 9　　　　　　　B. 10　　　　　　　C. 11　　　　　　　D. 12

答案	C
解析	【题干】：一个账户的存款为 ＄800，年复利为 2%，如果没有新的存款。 【问题】：多少年之后该账户的存款额为 ＄990？
巧解	根据题意得到该账户的存款额可以用一个方程来表示：$A = P(1+r)^t$，其中 $P=$ 账户的起始值 800，r 为年复利，为 0.02，t 是时间。 采用选项代入，一般可以从 B 项（中间值代入法）代入尝试，得到 $800\times(1.02)^{10}\approx 975$，小于 990，这个数值需要再增加。因此再用 C 项代入 $800\times(1.02)^{11}\approx 994.7$，符合条件，选 C。

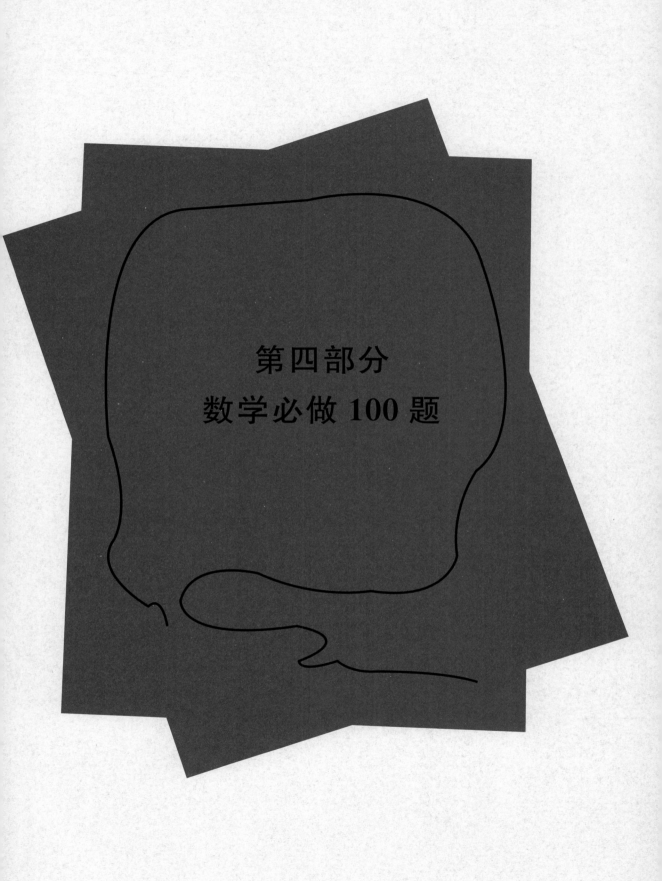

第四部分
数学必做 100 题

Practice 1

 1. In 1995, Diana read 10 English books and 7 French books. In 1996, she read twice as many French books as English books. If 60% of the books that she read during the 2 years were French, how many English and French books did she read in the 2 years?

A. 16　　　　B. 32　　　　C. 48　　　　D. 65

2.
$$C = \frac{5}{9}(F - 32)$$

The conversion of temperatures measured in F, degrees Fahrenheit, to C, degrees Celsius, is given by the above equation. Which of the following equations could be used to convert temperatures from degrees Celsius to degrees Fahrenheit?

A. $F = \frac{9}{5}C + 32$　　　　B. $F = \frac{9C - 32}{5}$

C. $F = \frac{9C}{5} - 32$　　　　D. $F = \frac{9C + 32}{5}$

 3. A college professor with several hundred students has office hours between classes to provide extra help when needed. His classes on Monday are from 9:00 AM to 10:45 AM and 2:30 PM to 3:45 PM. It takes him 5 minutes to walk from the classroom to his office, and he takes a lunch break from 12:00 PM to 1:00 PM. On a particular Monday, he plans to grade tests, which have all multiple-choice questions. If each test consists of 50 questions and it takes him 4 seconds to mark each question right or wrong, how many complete tests can he mark during his office hours if no students come for help? Assume that he does not take the time to add up the scores until after his afternoon class.

A. 46　　　　B. 47　　　　C. 54　　　　D. 55

 4. The concentration of a certain chemical in a full water tank depends on the depth of the water. At a depth that is x feet below the top of the tank, the concentration is $3 + \frac{4}{\sqrt{5 - x}}$ parts per million, where $0 < x < 4$. To the nearest 0.1 ft, at what depth is the concentration equal to 6 parts per million?

A. 2.4 ft　　　　B. 2.8 ft

C. 3.0 ft　　　　D. 3.2 ft

5. The outline of a sign for an ice-cream store is made by placing $\frac{3}{4}$ of the circumference of a circle with radius 2 feet on top of an isosceles triangle with height 5 feet, as shown (Fig. 4-1). What is the perimeter, in feet, of the sign?

A. $3\pi + 3\sqrt{3}$
B. $3\pi + 6\sqrt{3}$
C. $3\pi + 2\sqrt{3}$
D. $4\pi + 3\sqrt{3}$

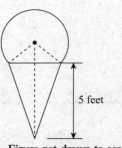

Note: Figure not drawn to scale

Fig. 4-1

6.
$$rx^2 = \frac{1}{s}x + 3$$

A quadratic equation is provided above, where r and s are constants. What are the solutions for x?

A. $x = \dfrac{1}{2sr} \pm \dfrac{\sqrt{\dfrac{1}{s^2} + 12r}}{2r}$

B. $x = \dfrac{1}{2sr} \pm \dfrac{\sqrt{-\dfrac{1}{s^2} - 12r}}{2sr}$

C. $x = \dfrac{s}{2r} \pm \dfrac{\sqrt{\dfrac{1}{s^2} - 12r}}{2r}$

D. $x = \dfrac{s}{2r} \pm \dfrac{\sqrt{s^2 - 12sr}}{2sr}$

7.
$$n = 12 \times 2^{\frac{t}{3}}$$

The number of mice in a certain colony is shown by the formula above, such that n is the number of mice and t is the time, in months, since the start of the colony. If 2 years have passed since the start of the colony, how many mice does the colony contain now?

8. In the figure (Fig. 4-2), AB is the arc of the circle with center O. Point A lies on the graph of $y = x^2 - b$, where b is a constant. If the area of shaded region AOB is π, then what is the value of b?

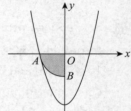

Fig. 4-2

9. Which of the following equations has a vertex of $(3, -3)$?

A. $y = 5(x - 3)^2 - 3$
B. $y = 5(x + 3)^2 - 3$
C. $y = 5(x - 3)^2 + 3$
D. $y = 5(x + 3)^2 + 3$

10. If $9^{-2} = \left(\dfrac{1}{3}\right)^x$, what is the value of x?

A. 1   B. 2   C. 4   D. 6

11. If the expression $\dfrac{9x^2}{3x - 2}$ is rewritten in the equivalent form $\dfrac{4}{3x - 2} + A$, what is A in terms of x?

A. $\dfrac{9x^2}{4}$ B. $3x + 2$ C. $3x - 2$ D. $9x^2 - 4$

12. A merchant sells three types of clocks that chime as indicated by the check marks in the table (Table 4 – 1). What is the total number of chimes of the inventory of clocks in the 90-minute period from 7:15 to 8:45?

Table 4 – 1   Inventory of Clocks and Frequency of Chimes

	Number of Clocks	Chimes n Times on the nth Hour	Chimes Once on the Hour	Chimes Once on the Half Hour
Type A	10	√		√
Type B	5	√		
Type C	3		√	√

 13. The amount of pancake mix required to make pancakes is proportional to the number of pancakes that are being made. The table (Table 4 – 2) shows the required amount of pancake mix and water to make 6 pancakes. How many cups of water are needed to make 15 pancakes?

Table 4 – 2

Pancake	Amount of Mix	Amount of Water
6	1 cup	$\frac{3}{4}$ cup

A. $1\frac{1}{4}$ B. $1\frac{3}{8}$

C. $1\frac{5}{8}$ D. $1\frac{7}{8}$

 14. Which of the following gives the length of chord DF in the figure (Fig. 4 – 3)?

A. 2cos 1.7
B. 2sin 1.7
C. 4cos 0.85
D. 4sin 0.85

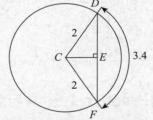

Note: Figure not drawn to scale

Fig. 4 – 3

 15. A company has two manufacturing plants, Plant A and Plant B. Three times Plant A's production is equal to one half of the total produced by the company. If T represents the total produced by the company, which of the following is equal to the amount produced by Plant B?

A. $\frac{5}{6}T$ B. $\frac{2}{3}T$ C. $\frac{1}{2}T$ D. $\frac{1}{6}T$

16. If there are no gaps between tiles used for a kitchen floor, how many 8-square-inch tiles are needed to cover a kitchen floor that is 10 feet wide and 14 feet long?

   A. 12
   B. 18
   C. 210
   D. 2,520

17. The circle shown (Fig. 4-4) is given by the equation $x^2 + y^2 + 6x - 4y = 12$. What is the shortest distance from A to B?

   A. 5
   B. 10
   C. $4\sqrt{3}$
   D. 24

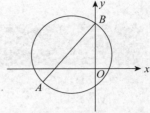

Note: Figure not drawn to scale

Fig. 4-4

18. Given the polynomial $6x^4 + 2x^2 - 8x - c$, where c is a constant, for what value of c will $\dfrac{6x^4 + 2x^2 - 8x - c}{x+2}$ have no remainder?

   A. -120 B. -60 C. 60 D. 120

19. If h is a function defined over the set of all real numbers and $h(x-4) = 6x^2 + 2x + 10$, then which of the following defines $h(x)$?

   A. $h(x) = 6x^2 - 2x + 114$
   B. $h(x) = 6x^2 - 46x + 98$
   C. $h(x) = 6x^2 + 2x + 98$
   D. $h(x) = 6x^2 + 50x + 114$

20. An online movie subscription service charges a dollars for the first month of membership and b dollars per month after that. If a customer has paid \$108.60 so far for the service, which of the following expressions represents the number of months he has subscribed to the service?

   A. $\dfrac{108.60}{a+b}$
   B. $\dfrac{108.60 - a}{b}$
   C. $\dfrac{108.60 - a - b}{b}$
   D. $\dfrac{108.60 - a + b}{b}$

■ Practice 1 答案与解析

1	D	【题干】：1995年Diana读了10本英文书和7本法文书。1996年她读的法文书是英文书的2倍。如果2年内她读的书中60%是法文书。 【问题】：她在两年中一共读了多少本英文书和法文书？ 【解析】：根据题意，可设1996年她读的法文书有F本，英文书有E本，1995年为：$10+7=17$本，1996年为：$F=2E$，可得$60\% = \dfrac{7+F}{17+E+F} = \dfrac{7+2E}{17+3E}$，解得$E=16$，$F=32$，因此总计为$17+16+32=65$，选D。

第四部分 数学必做100题

2	A	**【题干】**：温度 F（华氏度数）和温度 C（摄氏度数）之间的转换公式如上式所示。 **【问题】**：下列式子中哪一个式子表示的是从摄氏度数（C）转换为华氏度数（F）？ **【解析】**：根据题中的条件 $C = \frac{5}{9}(F-32)$ 进行转换可得 $\frac{9C}{5} = F - 32$，进一步整理可得 A。
3	A	**【题干】**：一个大学教授有上百名学生。在他的课间办公时间会给有需要的学生额外帮助。该教授在周一的课从上午 9 点到 10 点 45 以及下午 2 点 30 分到 3 点 45 分。从教室走到他的办公室需要 5 分钟。在 12 点到下午 1 点是该教授的午餐时间。某个周一该教授计划做一个等级测试。该测试为多项选择题。假设每份测试含 50 道题，每题教授需要用 4 秒来判断对错。 **【问题】**：如果没有学生需要帮助，则在该教授的工作时间内可以评阅多少份完整的测试（假设直到下午上课该教授都没有时间来加算成绩）？ **【解析】**：根据题意可得，该教授可用的时间为除上午 9 点到 10 点 45 以及下午 2 点 30 分到 3 点 45 分外除去午餐时间，共计 2 小时 45 分钟。注意还要扣除往返教室与办公室的时间（5 分钟×2 = 10 分钟）。所以实际可用的时间为 2 小时 45 分钟 − 10 分钟 = 2 小时 35 分钟 = 155 分钟 = 9,300 秒。每题判断需要 4 秒，一份测试含有 50 道题，所以共计评阅了 $\frac{9,300}{50 \times 4} = 46.5$ 份测试，注意由于"46.5"的小数位"0.5"意味着第 47 份测试还没有评阅完。所以完成的评阅只有 46 份，选 A。
4	D	**【题干】**：一个水箱中某化学物质的浓度与水的深度有关，在低于水箱顶部 $x(0<x<4)$ 英尺的深度，其浓度为百万分之 $3 + \frac{4}{\sqrt{5-x}}$。 **【问题】**：在多深的位置该化学物质的浓度为百万分之 6（近似到 0.1 英尺）？ **【解析】**：根据题意可得 $3 + \frac{4}{\sqrt{5-x}} = 6$，反求 x，解得 $x = 3.22222$，题目要求近似到 0.1，则 $x \approx 3.2$，选 D。
5	B	**【题干】**：图中给出的是一个冰淇淋店牌子的外形，其图案由一个半径为 2 英尺圆的 $\frac{3}{4}$ 部分放在一个高为 5 英尺的等腰三角形上构成。 **【问题】**：该牌子外形的周长是多少？ **【解析】**：根据题意可得，本题的关键是要求解出该等腰三角形腰的长度。根据圆的 $\frac{3}{4}$ 这个条件可以知道：以圆半径为腰的那个小等腰三角形第三条边对应的圆心角为 $360° \times \frac{1}{4} = 90°$，说明该小等腰三角形为直角等腰三角形，则可以得到该三角形的底边长为 $2\sqrt{2}$，再根据勾股定理得到大等腰三角形的腰长为 $3\sqrt{3}$，所以该牌子外形的周长为：$\frac{3}{4} \times 4\pi + 2 \times 3\sqrt{3} = 3\pi + 6\sqrt{3}$，选 B。
6	A	**【题干】**：给出了一个二次方程式，其中 r 和 s 都是常数。 **【问题】**：x 的解为多少？ **【解析】**：根据题意可将原式化简为：$rx^2 - \frac{1}{s}x - 3 = 0$，再根据维达定理可得 $x = \frac{-b \pm \sqrt{b^2 - 4ac}}{2a}$，选 A。

7	3,072	【题干】：一个克隆组中小鼠的数量如该公式所示，其中 n 代表小鼠的数量，t 代表从克隆开始以来的时间，以月来计算。 【问题】：从克隆开始两年至今，一共生了多少只小鼠？ 【解析】：根据题意，本题需要首先进行不同时间单位间的换算。公式中 t 是以月来计算的，因此要把两年换算成 $2\times12=24$ 个月，之后再代入到公式中可得：$n=12\times2^{\frac{24}{3}}=3,072$。
8	4	【题干】：如图所示，AB 是以 O 为圆心的圆弧，点 A 在图像 $y=x^2-b$ 上，b 是一个常数，如果阴影部分 AOB 的面积为 π。 【问题】：b 的值为多少？ 【解析】：根据题意可得，阴影部分 AOB 是四分之一圆。所以以 O 为圆心的圆面积为 4π，$4\pi=\pi r^2$，解得 $r=2$。说明点 A 的坐标为 $(-2,0)$，由于该点也在 $y=x^2-b$ 上，因此将 $x=-2$，$y=0$ 代入 $y=x^2-b$，可得 $b=4$。
9	A	根据题意得，要满足函数有一个 $(3,-3)$ 的顶点，说明当 $x=3$ 时，$y=-3$，能满足这一条件的为 A 项。
10	C	根据题意得：$\frac{1}{9^2}=\left(\frac{1}{3}\right)^4=\left(\frac{1}{3}\right)^x$，$x=4$，选 C。
11	B	【题干】：如果表达式 $\frac{9x^2}{3x-2}$ 可以改写成 $\frac{4}{3x-2}+A$ 的形式。 【问题】：用 x 来表示 A 该如何表示？ 【解析】：根据题意可得 $A=\frac{9x^2}{3x-2}-\frac{4}{3x-2}=\frac{9x^2-4}{3x-2}=\frac{(3x+2)(3x-2)}{3x-2}=3x+2$，选 B。
12	149	【题干】：一个商人卖三种不同类型的报时钟，钟的库存量和鸣钟信息如表所示。每台钟各自鸣响，A 型钟共有 10 台，整点鸣相应的声响（如 7 点钟鸣 7 下），半点钟鸣 1 下。B 型钟共有 5 台，整点鸣相应的声响（如 7 点钟鸣 7 下）。C 型钟共有 3 台，整点鸣 1 下，半点钟鸣 1 下。 【问题】：在 7:15 到 8:45 这 90 分钟内库存的钟一共鸣几下？ 【解析】：根据题意，在 90 分钟内钟鸣响的时间点有 3 处： 7:30：A 型钟鸣 1 下，10 台钟共计鸣 10 下；C 型钟鸣 1 下，3 台钟共计鸣 3 下，总计鸣 13 下。 8:30：同理也鸣 13 下。 8:00：A 型钟鸣 8 下，10 台钟共计鸣 80 下；B 型钟鸣 8 下，5 台钟共计鸣 40 下；C 型钟鸣 1 下，3 台钟共计鸣 3 下。 总计鸣：$13+80+40+3+13=149$。
13	D	【题干】：需要按照一定比例的煎饼粉来制作薄烤饼。煎饼粉、水与薄烤饼数量的比例如表所示。 【问题】：现在要制作 15 张薄烤饼，需要多少杯水？ 【解析】：根据题意设需要 w 杯水，因此可以得到一个对应的比例：$\frac{\frac{3}{4}}{6}=\frac{w}{15}$，解得 $w=1\frac{7}{8}$，选 D。
14	D	【题干】：如图所示，圆心为 C 的圆，弧 DF 长度为 3.4，CD 和 CF 是圆的半径，长度为 2，CE 与 DF 垂直。 【问题】：求弦 DF 的长度为多少？ 【解析】：根据题意可得，弧 DF 长 $L=\angle DCF\times CD$，即 $3.4=\angle DCF\times 2$，得到 $\angle DCF=1.7$。CD 和 CF 相等，且 CE 与 DF 垂直，所以 CE 平分 $\angle DCF$，所以 $\angle DCE=\frac{\angle DCF}{2}=0.85$。在直角三角形中，$\sin\angle DCE=\frac{DE}{CD}$，解得 DE 长度 $=\sin\angle DCE\times CD=2\sin 0.85$，因此得到 DF 长度 $=DE$ 长度 $\times 2=2\times 2\sin 0.85=4\sin 0.85$，选 D。

15	A	**【题干】**：一家公司有两个工厂 A 和 B，A 工厂产量的 3 倍是公司总产量的一半，如果 T 代表公司的总产量。 **【问题】**：下面哪个式子代表的是 B 工厂产量？ **【解析】**：根据题意可得，设 A 工厂的产量为 A，B 工厂的产量为 B，可以得到 $3A = \frac{1}{2}T$，解得 $A = \frac{1}{6}T$，因此 $B = T - \frac{1}{6}T = \frac{5}{6}T$，选 A。
16	D	**【题干】**：假设给厨房地面贴瓷砖，瓷砖之间没有缝隙。如果一个厨房的地面为 10 英尺宽和 14 英尺长。 **【问题】**：需要面积为 8 平方英寸的瓷砖多少个？ **【解析】**：根据题意可得，需要将英尺与英寸之间进行转换后再计算，需要的瓷砖数为：$\frac{10 \times 12 \times 14 \times 12}{8} = 2,520$，选 D。
17	B	**【题干】**：如图所示，圆的方程为 $x^2 + y^2 + 6x - 4y = 12$。 **【问题】**：A 和 B 两点的最小距离为多少？ **【解析】**：根据题意可得，A 和 B 两点的最小距离即为该圆的直径。首先需要将圆的方程写成标准方程形式： $$x^2 + 6x + 9 + y^2 + 4y + 4 = 12 + 9 + 4,$$ 变换形式后为 $(x+3)^2 + (y+2)^2 = 5^2$，可以得到圆的半径为 5，A 和 B 两点的最小距离即为该圆的直径，为 10，选 B。
18	D	**【题干】**：给出一个多项式 $6x^4 + 2x^2 - 8x - c$，c 是一个常数。 **【问题】**：当 c 的值为多少时 $\frac{6x^4 + 2x^2 - 8x - c}{x+2}$ 没有余数？ **【解析】**：根据题意可得，$\frac{6x^4 + 2x^2 - 8x - c}{x+2}$ 没有余数说明 $x+2$ 是 $6x^4 + 2x^2 - 8x - c$ 的一个因式。根据因式定理，将 $x = -2$ 代入式子 $6x^4 + 2x^2 - 8x = c$ 中，得到 $c = 120$，选 D。
19	D	**【题干】**：h 是一个函数，满足所有实数，且 $h(x-4) = 6x^2 + 2x + 10$。 **【问题】**：$h(x)$ 可以定义为下面哪一个式子？ **【解析】**：根据题意需要做一下代数转换，设 $a = x - 4$，所以 $x = a + 4$，代入 $h(a)$，得到 $h(a) = 6(a+4)^2 + 2(a+4) + 10 = 6a^2 + 50a + 114$，$h$ 是一个函数，满足所有实数，所以 $h(x) = 6x^2 + 50x + 114$，选 D。
20	D	**【题干】**：一个网上电影订阅服务向其会员第 1 个月收费 $\$a$，之后每个月收取 $\$b$，如果消费者共计花费 $\$108.60$ 购买该项订阅服务。 **【问题】**：下面哪一个式子表达的是该消费者订阅服务的月数？ **【解析】**：根据题意，可以设订阅服务的月数为 m，因此收费就分为了两部分：第一个月为 a，从第二个月开始一共 $(m-1)$ 个月收费 $b \times (m-1)$，共计收费 $a + (m-1) \times b = 108.6$，将该式子整理后用 a 和 b 来表达 m 为 $\frac{108.60 - a + b}{b}$，选 D。

Practice 2

1. The average acceleration of a car in miles per hour² over a period of h hours is calculated using the formula $a = \frac{f-s}{h}$, where s is the speed in miles per hour at the beginning of the period and f is the speed in miles per hour at the end of the period. What is s in terms of f, a, and h?

   A. $s = \frac{ah}{f}$ B. $s = ah - f$ C. $s = f - ah$ D. $s = a - fh$

2. How many integers satisfy the inequality $|x| < \pi$?

   A. 3   B. 4   C. 7   D. more than 7

3. At Joe's Pizzeria, small pizzas cost \$7.50 and large pizzas cost \$11.00. One day from 3:00 P.M. to 9:00 P.M., Joe sold 100 pizzas and took in \$848. Solving which of the following systems of equations could be used to determine the number of small pizzas, S, and the number of large pizzas, L, that Joe sold during that 6-hour period?

   A. $S + L = 848$, $7.5S + 11L = 100$ B. $S + L = 848$, $7.5S + 11L = \frac{848}{6}$

   C. $S + L = 100$, $7.5S + 11L = 848$ D. $S + L = 100$, $7.5S + 11L = 848 \times 6$

4. In the figure (Fig. 4-5), two searchlights S_1 and S_2 are located 10,000 feet apart, each covers an area of radius 10,000 feet, and each is located 8,000 feet from the railroad track. To the nearest 1,000 feet, what is the total length x of track spanned by the searchlights?

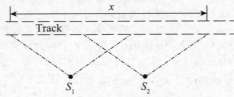

Note: Figure not drawn to scale

Fig. 4-5

5. A certain company has records stored with a record-storage firm in 15-inch by 12-inch by 10-inch boxes. The boxes occupy 1.08 million cubic inches of space. If the company pays \$0.25 per box per month for record storage, what is the total amount that the company pays each month for record storage?

A. $150 B. $300
C. $600 D. $1,200

6. To celebrate a colleague's retirement, the coworkers in an office agree to contribute equally to a catered lunch that costs a total of b dollars. If there are a coworkers in the office, and if c coworkers fail to contribute, which of the following represents the extra amount in dollars, that each of the remaining coworkers must contribute to cover the cost of lunch?

A. $\dfrac{b}{c}$ B. $\dfrac{b}{a-c}$

C. $\dfrac{bc}{a-c}$ D. $\dfrac{bc}{a(a-c)}$

7. If $-1 \leqslant a \leqslant 2$, $-3 \leqslant b \leqslant 2$, what is the greatest possible value of $(a+b)(b-a)$?

8. If each number in the following sum were increased by t, the new sum would be 4.22. What is the value of t?

$$\begin{array}{r} 0.65 \\ 0.85 \\ 0.38 \\ +0.86 \\ \hline 2.74 \end{array}$$

A. 0.24 B. 0.29 C. 0.33 D. 0.37

9. A store charges $39 per pair for a certain type of pants. This price is 30% more than the wholesale price. At a Thanksgiving sale, store employees can purchase any remaining items at 40% off the wholesale price. How much would it cost an employee to purchase a pair of pants of this type at this sale?

A. $12.00 B. $14.00
C. $18.00 D. $21.00

Question 10 and 11 refer to the following information.

10. Karen runs a flower shop. She determines that it takes her two hours of online marketing to bring in five new orders. If each order bills an average of $30, how many hours of marketing are necessary for her business to bill $10,000 a month?

11. Karen hires a marketing assistant to bolster her online presence, and finds that it now takes only one hour of online marketing to bring in five new orders. If she pays her assistant $15 per hour, and the cost to fill an order is $5, how many hours must her assistant work each month for Karen's business to make a monthly profit of $10,000?

12. Priya is buying a used car. She has narrowed her search down to two cars. The following table (Table 4-3) shows the information about each car.

Table 4-3

Car	Purchase price	Gas Mileage (miles per gallon)	Tax, Registration, and Repairs
A	$3,500	20	$800
B	$5,000	25	$400

Priya likes Car B better, because it is newer and gets better gas mileage, but she has calculated that it will cost her less to buy and use Car A based on the following criteria:
- Priya estimates that she will drive approximately 400 miles per month.
- The average cost of gasoline per gallon for her area is $3.80.
- Priya plans on owning the car for 3 years.

Based on the data, how much less will it cost Priya to buy and use Car A?
A. $552.80   B. $952.80   C. $1,054.40   D. $1,084.80

13. If $a + bi$ represents the complex number that results from multiplying $3 + 2i$ by $5 - i$, what is the value of a?
A. 2   B. 13   C. 15   D. 17

14. Three couples host a dinner party. Each couple invites 4 guests, none of whom are the same. A table for the party seat 5 people. If everyone attends the party, what is the minimum number of tables needed to seat everyone?

15. If $(i^{413})(i^x) = 1$, then what is one possible value of x?
A. 0   B. 1   C. 2   D. 3

16. Trains A, B, and C passed through a station at different speeds. Train A's speed was 3 times Train B's speed, and Train C's speed was twice Train A's. What was Train C's speed, in miles per hour, if Train B's speed was 7 miles per hour?

17. If a quadratic equation is used to model the data shown in the scatterplot (Fig. 4-6), and the model fits the data exactly, which of the following is a solution to the quadratic equation?
A. 28   B. 32   C. 34   D. 36

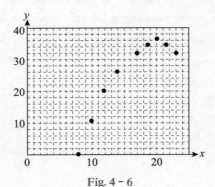

Fig. 4-6

18. Generally, the price of an item is a good indicator of how many units of that item will be sold. The lower the price, the more units will be sold. A marketing department develops a table (Table 4 - 4) showing various price points and the projected number of units sold at that price point. Which of the following represents the linear relationship shown in the table, where x is the price and y is the number of units sold?

Table 4 - 4

Price Per Pencil	Projected Number of Units Sold
$ 0.20	150,000
$ 0.25	135,000
$ 0.30	120,000
$ 0.35	105,000
$ 0.40	90,000
$ 0.45	75,000

A. $y = 0.03x + 150,000$ B. $y = -300,000x + 75,000$
C. $y = -300,000x + 900,000$ D. $y = -300,000x + 210,000$

 19. A survey of college freshmen asked them if they have decided on a major. Part of the information that was gathered is shown in the table (Table 4 - 5). What percent of the women surveyed said that they have chosen a major?

Table 4 - 5

	Men	Women	Total
Yes	1,000		
No		500	
Total	3,000		5,000

20. A circle has a radius of 8. A cone is formed by cutting out a 60° wedge and placing together the two radii of the shaded part of the circle (Fig. 4 - 7). What is the surface area of the cone rounded to the nearest whole number, excluding the circular base?

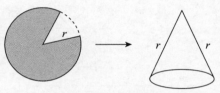

Note: Figure not drawn to scale

Fig. 4 - 7

■ Practice 2 答案与解析

1	C	【题干】：一辆车在一个时间段 h 小时内的平均加速度可以用以下的公式表示，$a = \frac{f-s}{h}$，其中 s 代表开始时的车速，f 代表该时间段结束时的车速。 【问题】：如何用 f，a 和 h 来表示 s？ 【解析】：根据题意由 $a = \frac{f-s}{h}$ 可以转换为 $ah = f - s$，再变形可以得到 $s = f - ah$，选 C。
2	C	【题干与问题】：有多少个整数满足 $\lvert x \rvert < \pi$？ 【解析】：根据题意得到： $\lvert x \rvert < \pi \Rightarrow -\pi < x < \pi \Rightarrow -3.14 < x < 3.14$ 因此一共有七个整数符合条件：-3，-2，-1，0，1，2，3，选 C。
3	C	【题干】：Joe 的比萨店出售两种比萨，小尺寸比萨 \$7.50 一个，大尺寸比萨 \$11.00 一个。从下午 3 点到晚上 9 点 6 个小时内共卖出 100 份比萨，收入为 \$848。 【问题】：用 S 代表小尺寸比萨，用 L 代表大尺寸比萨，下面哪一个方程组符合题目条件描述？ 【解析】：根据题意可得： 数量关系：$S + L = 100$ 价钱关系：$7.50S + 11.00L = 848$，选 C。
4	22,000	【题干】：在给出的图形中，两个探照灯 S_1 和 S_2 相隔距离为 10,000 英尺，每个探照灯能照射的范围是一个半径为 10,000 英尺范围的圆，并且两个探照灯距离铁路铁轨的距离为 8,000 英尺。 【问题】：探照灯能照射到的铁轨总长度是多少？ 【解析】：根据题意可以得到以下的图形（图 4-1）：从 S_1 和 S_2 分别向铁轨做垂线得到 S_1B 和 S_2C，则 $S_1B = S_2C = 8,000$，$BC = S_1S_2 = 10,000$。在直角三角形 AS_1B 和 S_2CD 中 $AB = CD = \sqrt{10,000^2 - 8,000^2} = 6,000$，所以能够照射到的长度为 $AD = AB + BC + CD = 6,000 + 10,000 + 6,000 = 22,000$。 图 4-1

5	A	【题干】：某公司的唱片被唱片保管公司存放在体积为 15 英尺×12 英尺×10 英尺的盒子中，这些盒子占用了 108 万立方英尺的空间。该公司每月给一个盒子支付 $0.25 的保管费用。 【问题】：该公司每月需要支付的唱片保管费共计为多少？ 【解析】：根据题意，先计算出需要保管的唱片总盒数：$\frac{1,080,000}{15 \times 12 \times 10} = 600$。 再计算总价为 $0.25×600 = $150，选 A。
6	D	【题干】：为了庆祝一个同事的退休，同一个办公室的其他同事决定以平摊的形式搞一个午餐会，该午餐会的总费用为 $b。现在一共有 a 位同事，其中 c 位同事没有参与平摊。 【问题】：下列哪一个式子代表剩余的同事为该午餐会多平摊的那部分钱？ 【解析】：根据题意可得，要得到多平摊的那部分，需要首先获得每个人正常平摊和非正常平摊的费用。 正常情况下，总价 $b，有 a 位同事，每人平摊 $$\frac{b}{a}$$，非正常情况下，由于 c 位同事不参与平摊，实际人数为 $a-c$ 人，每人平摊 $$\frac{b}{a-c}$$。这样可以得到两部分的差为 $\frac{b}{a-c} - \frac{b}{a}$，即 $\frac{ab-b(a-c)}{(a-c)a} = \frac{bc}{a(a-c)}$，选 D。
7	9	【题干与问题】：如果 $-1 \leq a \leq 2$，$-3 \leq b \leq 2$，则式子 $(a+b)(b-a)$ 的最大值为多少？ 【解析】：根据题意可得：$(a+b)\times(b-a) = b^2 - a^2$，可见要想该式的值最大，则 b 要取最大值，a 取最小值。从题中的 a 和 b 的取值范围来看，a^2 最小为 0（当 $a=0$ 时），b^2 最大可以取 9（当 $b=3$）时，因此 $b^2 - a^2$ 的最大值为 9。
8	D	【题干】：下列求和的每个数都增加了 t，则得到新的和为 4.22。 【问题】：t 的值为多少？ 【解析】：根据题意，原式中的和为 2.74，增加后为 4.22，因此共计增加了 4.22 - 2.74 = 1.48。由于每个数都增加了 t，一共增加了 4t，所以 $t = \frac{1.48}{4} = 0.37$，选 D。
9	C	【题干】：一家商店卖一种款型的裤子，价格为每条 $39。这个价格比批发价多 30%。在感恩节促销活动中，店员购买裤子按其批发价再打 6 折。 【问题】：在促销活动中这种款式的裤子卖了多少钱？ 【解析】：根据题意可得，设这种款式的裤子批发价为 $w，可得 $w\times(1+30\%) = 39$，解得 $w = 30$。再按 6 折计算，可得 30×60% = $18，选 C。
10	134	【题干】：Karen 开了一家花店，她发现在网上商城需要花 2 小时处理 5 个新的订单，每个订单均价为 $30。 【问题】：她一个月需要在网上商城工作多长时间才能保证有 $10,000 的收入？ 【解析】：每小时的收入：$\frac{30\times 5}{2} = \$75$，一个月需要工作的时间：$\frac{10,000}{75} = 133.33 \approx 134$ （注意本题问的是需要多长时间才能达到某个值，因此需根据逻辑含义对数据的小数进行处理。）

11	91	【题干】：她雇了1位市场助理来帮助她打理网上商城的业务，发现现在只需要1小时就可以处理五个新订单。她要支付每小时 $15 给她的助理，一个订单的成本是 $5。 【问题】：她的助理需要工作多长时间才能满足 Karen 每个月的利润是 $10,000？ 【解析】：根据题意可得：利润＝收入－成本。本题中每小时的成本是：五个订单的成本＋助理的工资 ＝ $5×5＋$15 ＝ $40，每小时的收入为 $30×5 ＝ $150，所以每小时的利润为 $150－$40 ＝ $110，那么一共要获得 $10,000 的利润，需要的小时数为：$\frac{10,000}{110}$ ＝ 90.909≈91。
12	A	【题干】：Priya 想买一辆二手车，她最后倾向于两辆车。表中给出了这两辆车的有关信息。表中第一列为车的价格，第二列为油耗（英里/加仑），第三列为交税、注册以及维修等费用。Priya 更中意 B 车，因为该车更新、更省油，但是经过计算她认为应该买 A 车，理由如下： 1) Priya 估计她每月大约开 400 英里；2) 她所在区域油价为每加仑 $3.80；3) 她想拥有这辆车 3 年。 【问题】：基于以上信息，Priya 买 A 车将少花多少钱？ 【解析】：根据题意，买车的总体花费包括了车价、油费以及交税、注册和修理的费用。注意其中有两个单位换算：一个是每月开车的里程与油耗之间的对应与计算，油费计算为油价/油耗（英里/加仑）×3 年的行驶总里程；第二就是开车的月里程与总开车时间（以年为单位）的对应与计算。 购买 A 车的花费为：$3,500 ＋ $400×36×$\frac{$3.8}{20}$ ＋ $800 ＝ $7,036； 购买 B 车的花费为：$5,000 ＋ $400×36×$\frac{$3.8}{25}$ ＋ $400 ＝ $7,588.8。 因此买 A 车便宜 $7,588.8－$7,036 ＝ $552.8，选 A。
13	D	【题干】：$a+bi$ 代表一个复数，等于 $3+2i$ 乘以 $5-i$。 【问题】：则 a 的值是多少？ 【解析】：$a+bi=(3+2i)×(5-i)=15-3i+10i-2i^2=17+7i$，$a=17$，选 D。
14	4	【题干】：三对夫妻办一个餐会，每一对夫妻邀请四位客人，邀请的客人都不重复。一张桌子可以坐五个人，且每个人都参加了餐会。 【问题】：每个人都坐到桌子上，则需要几张桌子？ 【解析】：根据题意可得：客人一共 3×4＝12 人，加上三对夫妻一共是 12＋2×3＝18 人，需要的桌数为 $\frac{18}{5}$＝3.6≈4（注意是桌数，这里不能是逻辑进位）。
15	D	【题干与问题】：如果 $(i^{413})(i^x)=1$，x 的一个可能值为多少？ 【解析】：根据题意可得，$(i^{413})(i^x)=i^{413+x}$，由于 $i^4=1$，因此 413＋x 必须是 4 的倍数，通过对四个选项判断后发现，当 $x=3$ 时，$\frac{416}{4}$＝104，符合上述条件，因此选 D。
16	42	【题干】：A，B 和 C 三辆火车以不同速度通过车站，A 车的速度是 B 车的 3 倍，C 车的速度是 A 车的 2 倍，B 车的速度为 7 英里/小时。 【问题】：C 车的速度是多少？ 【解析】：根据题意，设 A，B 和 C 车的速度分别为 A，B 和 C，可得：$A=3B$，$C=2A$，因为 $B=7$，所以 $A=21$，$C=2A=42$。

17	B	【题干】：一个二次方程表征的数据散点图如图所示，该函数与散点图的趋势完全匹配。 【问题】：下面哪一个是该二次方程的一个解？ 【解析】：根据题意可得，该二次方程图像是一个开口向下的抛物线，对称轴为 $x=20$，且与 x 轴的一个交点为（8, 0）。该交点与对称轴的距离为 $20-8=12$，且在对称轴的左侧。根据二次方程的对称性，该抛物线与 x 轴的另一个交点与对称轴的距离也为 12，且在对称轴的右侧。因此该交点为（32, 0）。抛物线的图像与 x 轴的交点即为该二次方程的解，即 $x=32$，选 B。
18	D	【题干】：一般来说，一个物品的价格是指示该物品到底能卖出多少量的指标。价格越低，卖出的量越大。一个公司的市场部门给出了一个表格，显示的是物品不同价格对应的销售量。 【问题】：下面哪一个式子表示该表格中的线性关系？其中 x 代表价格，y 代表销售量。 【解析】：根据题意可得，线性关系 $y=ax+b$ 中，斜率 a 代表的是 y 的变化量与 x 的变化量之间的比值。可将表格中的价格看作是自变量 x，销售量看作是应变量 y，从本题表格中找出两组 x，y 的值（0.2, 150,000）和（0.4, 90,000）： $a=\dfrac{150,000-90,000}{0.2-0.4}=-300,000$，可得 $y=-300,000x+b$，再将（0.2, 150,000）代入，得到 $y=-300,000x+210,000$，选 D。
19	75	【题干】：一项调查针对学院大一新生，问他们是否已经选择了专业，部分信息如表所示。 【问题】：女学生中已经选专业的所占百分比是多少？ 【解析】：根据题意将表中的部分信息补全：1）女学生总人数为 5,000−3,000=2,000 人；2）其中已经选专业的为 2,000−500=1,500 人。因此女学生中已经选专业的百分比为 $\dfrac{1,500}{2,000}\times 100\%=75\%$。
20	168	【题干】：如图所示，一个圆的半径为 8，从圆的边缘剪开一个 60°角，并将阴影部分的两边并在一起形成一个圆锥。 【问题】：问该圆锥的表面积为多少？（结果保留整数，圆锥表面积不包括圆锥的底面积） 【解析】：该圆锥的表面是由圆去掉一个圆心角为 60°的部分组成，即图上的阴影面积为：$\dfrac{360-60}{360}\times$ 完整圆面积 $=\dfrac{360-60}{360}\times\pi r^2=167.55\approx 168$。

Practice 3

1. $$s = -16t^2 + vt + c$$
   A ball is at an initial height of c feet. The equation above gives the height s, in feet, of a ball t seconds after it is thrown vertically in the air with an initial speed of v feet per second. Which of the following gives v in terms of s, t, and c?
   A. $v = \dfrac{s-c+16}{t}$
   B. $v = \dfrac{s-c}{t} + 16t$
   C. $v = \dfrac{s+c}{t} - 16t$
   D. $v = h + c - 16t$

2. A certain used-book dealer sells paperback books at 3 times dealer's cost and hardback books at 4 times dealer's cost. Last week the dealer sold a total of 120 books, each of which had cost the dealer $1. If the gross profit (sales revenue minus dealer's cost) on the sale of all of these books was $300, how many of the books sold were hardback?
   A. 40   B. 60   C. 75   D. 90

3. Which of the following inequalities is an algebraic expression for the shaded part of the number line (Fig. 4-8)?
   A. $|x| \leqslant 3$
   B. $|x| \leqslant 5$
   C. $|x-2| \leqslant 3$
   D. $|x+1| \leqslant 4$

   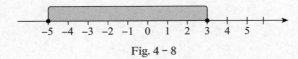

   Fig. 4-8

4. If each side of $\triangle ACD$ (Fig. 4-9) has length 3 and if AB has length 1, what is the area of region $BCDF$, in units squared?
   A. $\dfrac{9}{4}$
   B. $\dfrac{7}{4}\sqrt{3}$
   C. $\dfrac{9}{4}\sqrt{3}$
   D. $\dfrac{7}{2}\sqrt{3}$

   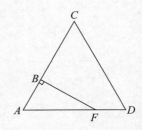

   Note: Figure not drawn to scale
   Fig. 4-9

5. A ladder 25 feet long is leaning against a wall that is perpendicular to level ground. The bottom of the ladder is 7 feet from the base of the wall. If the top of the ladder slips down 4 feet, how many feet will the bottom of the ladder slip?
   A. 4   B. 5   C. 8   D. 9

6. If $\sin x = \dfrac{1}{2}$ and x is between $\dfrac{\pi}{2}$ and $\dfrac{3\pi}{2}$, what is the value of $\dfrac{x}{2}$?
   A. $\dfrac{5\pi}{6}$ B. $\dfrac{7\pi}{12}$ C. $\dfrac{5\pi}{12}$ D. $\dfrac{\pi}{12}$

7. Carlos and Katherme are estimating acceleration by rolling a ball from rest down a ramp. At 1 second, the ball is moving at 5 meters per second (m/s); at 2 seconds, the ball is moving at 10 m/s; at 3 seconds, the ball is moving at 15 m/s; and at 4 seconds, it is moving at 20 m/s. When graphed on an xy-plane, which equation best describes the ball's estimated acceleration where y expresses speed and x expresses time?
   A. $y = 5x + 5$ B. $y = 25x$ C. $y = 5x$ D. $y = (4x+1)^2 + 5$

8. Of the 126 students who applied for a full scholarship to Kent College, 9 were successful. What is the ratio of students receiving a scholarship to those who are not?
   A. 1 to 11   B. 1 to 12   C. 1 to 13   D. 1 to 14

9. The angle created by an individual's line of vision from sea level to the top of a lighthouse is 60°. The lighthouse is known to rise 180 feet above sea level. What is the distance (to the nearest foot) between the observer and the base of the lighthouse?
   A. 104 feet   B. 180 feet   C. 208 feet   D. 254 feet

10. A locking pin is often made using a cylinder-cylinder pair in which a narrow cylinder fits tightly inside a wider cylinder. The inner cylinder protrudes from the outer cylinder, usually by equal amounts on both ends. In the following diagram (Fig. 4 – 10), the radius of the inner cylinder is half the radius of the outer cylinder, and it protrudes from the outer cylinder by 4 centimeters on each end.

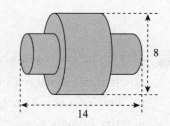

Note: Figure not drawn to scale

Fig. 4 – 10

What is the volume of the locking pin, to the nearest cubic centimeter?

11. A typical song downloaded from the Internet is 4 megabytes in size. Lindy has satellite Internet and her computer downloads music at a rate of 256 kilobytes per second. If 1 megabyte equals 1,024 kilobytes, about how many songs can Lindy download in 2 hours?
    A. 128 songs   B. 450 songs   C. 1,800 songs   D. 1,920 songs

12. The Venn diagram (Fig. 4 – 11) shows the distribution of 30 science students who studied butterflies, grasshoppers, both, or neither. What percent of the students studied butterflies only?

A. 10%   B. 20%   C. 30%   D. 40%

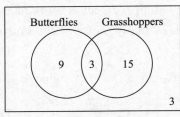

Fig. 4-11

13. At a beach of a rectangular swimming area with dimensions x and y meters and a total area of 4,000 square meters is marked off on three sides with rope. as shown in the figure (Fig. 4-12), and bounded on the fourth side by the beach. Additionally, rope is used to divide the area into three smaller rectangular sections. In terms of y, what is the total length, in meters, of the rope that is needed both to bound the three sides of the area and to divide it into sections?

A. $y + \dfrac{4{,}000}{y}$ B. $y + \dfrac{16{,}000}{y}$ C. $y + \dfrac{16{,}000}{3y}$ D. $3y + \dfrac{16{,}000}{3y}$

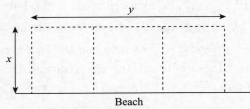

Note: Figure not drawn to scale

Fig. 4-12

14. One end of an 80-inch-long paper strip is shown in the figure (Fig. 4-13). The notched edge, shown in bold, was formed by removing an equilateral triangle from the end of each 4-inch length on one edge of the paper strip. What is the total length, in inches, of the bold notched edge on the 80-inch paper strip?

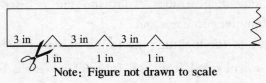

Note: Figure not drawn to scale

Fig. 4-13

15. A four-digit (0-9) password must meet the following restrictions:

• The first digit must be prime.
• The second digit must be odd.
• The third digit must he divisible by 5.
• The fourth digit must be a nonprime odd number.

How many possible passwords of this type exist?

A. 40 B. 45 C. 80 D. 90

16. $$\frac{\sqrt{x} \cdot x^{\frac{5}{6}} \cdot x}{\sqrt[3]{x}}$$

If x^n is the simplified form of the expression above, what is the value of n?

17. A salesperson's commission is k percent of the selling price of a car. Which of the following represents the commission, in dollars, on 2 cars that sold for $14,000 each?

A. $280k$ B. $7,000k$ C. $28,000k$ D. $\dfrac{14,000}{100+2k}$

18. The table (Table 4-6) shows the cost of the life span of several different types of light bulb. What is the ratio of the cost, in cents per hour of life span, of a compact florescent lamp to that of a light emitting diode?

A. 1.5 to 5 B. 3 to 8 C. 3 to 10 D. 4 to 5

Table 4-6　　　　　　　　Different Types of Light Bulbs

Type of Bulb	Cost	Life Span
Incandescent	$0.50	Approximately 1,000 hours
Compact Florescent Lamp (CFL)	$1.50	Approximately 10,000 hours
Light Emitting Diode (LED)	$10	Approximately 25,000 hours

19. The figure (Fig. 4-14) shows a fish tank with sand in the bottom. If the water level is to be 3 inches below the top, how many cubic inches of water are needed to fill the tank?

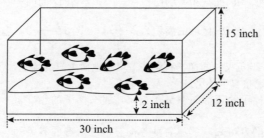

Note: Figure not drawn to scale
Fig. 4-14

20. The picture of a building is shown in a college architecture book. In the book, every half-inch represents 25 feet. How tall is the building in the book if the actual building is 350 feet tall?

A. 175 inches
B. 125 inches
C. 25 inches
D. 7 inches

■Practice 3 答案与解析

1	B	**【题干】**：一个球所在的原始高度为 c 英尺，下面的公式表示这个球以 v 英尺/秒的初速度竖直向上抛向高空 t 秒的高度 s。 **【问题】**：下面哪个式子用 s，t 和 c 来表示 v？ **【解析】**：根据题意，$s=-16t^2+vt+c$，转换为 $vt=s+16t^2-c$，左右两边同除以 t 后得到：$v=\dfrac{s-c}{t}+16t$，选 B。
2	B	**【题干】**：旧书商出售平装书的价格是成本价的三倍，出售精装书的价格是成本价的四倍。这个书商上周一共出售了 120 本书，每本书的成本为 \$1，如果出售书的毛利是 \$300。 **【问题】**：该书商出售了多少本精装书？ **【解析】**：gross profit 指毛利，毛利 = 销售收入 − 成本。根据题意设平装书的数量为 p，精装书的数量为 h，可得到方程组： $$p+h=120$$ $$3p+4y=300+120$$ 解得 $p=60$，$h=60$，该书商出售了 60 本精装书，选 B。
3	D	**【题干与问题】**：下面哪一个不等式表示的是图中数轴上的阴影部分？ **【解析】**：根据图中数轴得到：$-5\leqslant x\leqslant 3$，变形为 $-5+1\leqslant x+1\leqslant 3+1$，化简后得到 $\lvert x+1\rvert\leqslant 4$，选 D。
4	B	**【题干】**：$\triangle ACD$ 每条边长都是 3，且 AB 的长度为 1。 **【问题】**：区域 $BCDF$ 的面积是多少？ **【解析】**：因为 $\triangle ACD$ 是等边三角形，所以 $\angle A=60°$，$\triangle ABF$ 是直角三角形，所以 BF 长度 $=\sqrt{3}\times AB=\sqrt{3}$，因此 $\triangle ABF$ 的面积 $=\dfrac{1}{2}\sqrt{3}$；$\triangle ACD$ 的面积 $=\dfrac{1}{2}\times AC\times AD\times\sin 60°=\dfrac{9}{4}\sqrt{3}$。综上所述 $BCDF$ 区域的面积为：$\dfrac{9}{4}\sqrt{3}-\dfrac{1}{2}\sqrt{3}=\dfrac{7}{4}\sqrt{3}$，选 B。
5	C	**【题干】**：一个 25 英尺长的梯子斜靠在一堵与地面垂直的墙上，梯子的底部距墙根有 7 英尺。 **【问题】**：如果梯子的顶部向下滑 4 英尺，那么梯子的底部将滑多少英尺？ **【解析】**：根据题意可知，本题主要考查直角三角形的有关知识在实际生活中的应用。如图 4-2 所示： 在梯子没有滑动之前，梯子的垂直高度为 $\sqrt{25^2-7^2}=24$ 英尺，当梯子开始向下滑落 4 英尺后，梯子的垂直高度为 $24-4=20$ 英尺，则梯子距离墙根的距离为 $\sqrt{25^2-20^2}=15$，所以梯子的底部滑动了 $15-7=8$ 英尺，选 C。 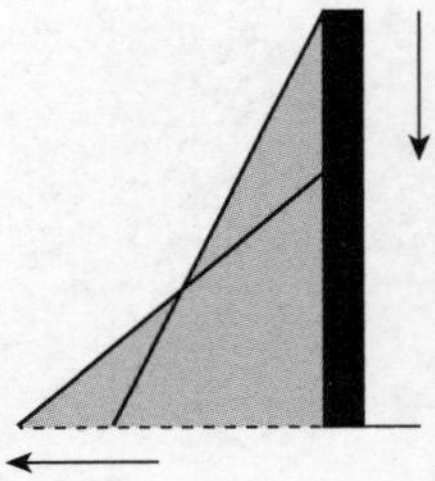 图 4-2

6	C	**【题干与问题】**：如果 $\sin x = \frac{1}{2}$，且 x 在 $\frac{\pi}{2}$ 到 $\frac{3\pi}{2}$ 之间，则 $\frac{x}{2}$ 的值是多少？ **【解析】**：根据题意可得 $\sin x = \frac{1}{2}$，且 x 的取值范围为 $\frac{\pi}{2}$ 至 $\frac{3\pi}{2}$，说明 x 在第二象限，如图 4-3 所示： 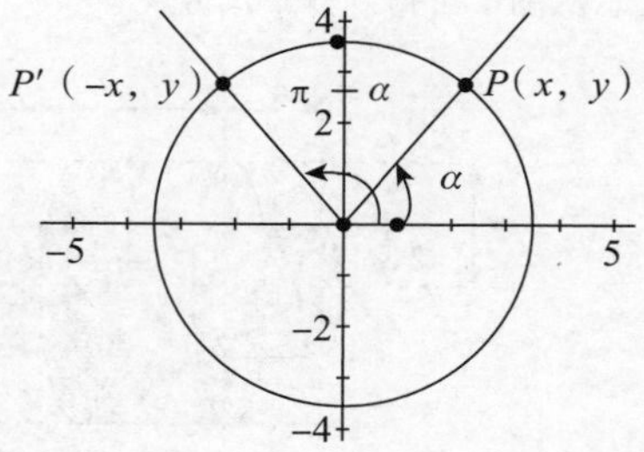 图 4-3 因此 $\sin x = \frac{1}{2} = \sin\left(\pi - \frac{\pi}{6}\right)$，解得 $x = \frac{5\pi}{6}$。注意此题问的是 $\frac{x}{2}$，所以答案为 $\frac{1}{2} \times \frac{5\pi}{6} = \frac{5\pi}{12}$，选 C。
7	C	**【题干】**：Carlos 和 Katherme 通过用一个球从斜坡上滚下来估算加速度。在第 1 秒，球每秒钟移动了 5 米。在第 2 秒，球每秒钟移动了 10 米。在第 3 秒，球每秒钟移动了 15 米。在第 4 秒，球每秒钟移动了 20 米。 **【问题】**：如果将该实验现象在 xy 坐标系中用图来表示，下列哪个式子最好地表述了球的加速度？其中 y 代表速度，x 代表时间。 **【解析】**：根据题意，在 xy 坐标系中用图来表示，则需要知道相应点的坐标。从原题中可以得到的数据，如果用不同的点 (x, y) 来表示，其中 y 代表速度，x 代表时间，则可以得到：$(0, 0)$、$(1, 5)$、$(2, 10)$、$(3, 15)$ 和 $(4, 20)$，一共五个点，可见 x 每增加 1，y 的值对应增加 5，说明斜率 $\text{slop} = \frac{\Delta y}{\Delta x} = 5$。根据该性质，满足条件的为 C 项。
8	C	**【题干】**：126 名学生申请 Kent 学院的奖学金，其中有 9 人成功。 **【问题】**：获得奖学金与未获得奖学金同学的比例是多少？ **【解析】**：根据题意可得获得奖学金的有 9 人，未获得奖学金的有 126-9 人，比例 = $\frac{9}{126-9} = \frac{1}{13}$，选 C。
9	A	**【题干】**：一个人从海平面看一座灯塔，需要仰起视角 60 度，这座灯塔离海平面 180 英尺高。 **【问题】**：该灯塔离观察者有多远？（结果保留到整数） **【解析】**：根据题意，设该灯塔离观察者的距离为 x，如图 4-4 所示，可得：$\tan 60° = \frac{180}{x} = \sqrt{3}$，解得 $x = 104$。选 A。 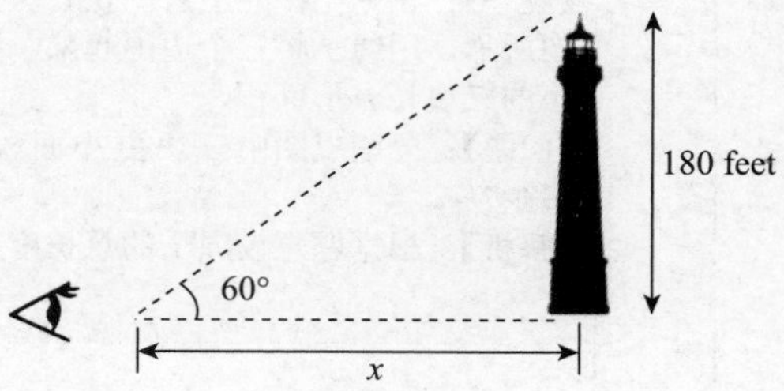 图 4-4

10	402	【题干】：一个锁插针通常由一对圆柱体组成：一个较窄的圆柱体紧紧插入到一个较宽的圆柱体中。如图4-5所示，靠内的那个圆柱体从靠外的那个圆柱体两端伸出且伸出的长度相等。靠内的圆柱体半径为靠外的圆柱体半径的一半，且两端伸出的长度为4厘米。 【问题】：该锁插针的体积是多少？（结果保留到整数） 【解析】：根据题意从图中可得到以下尺寸： 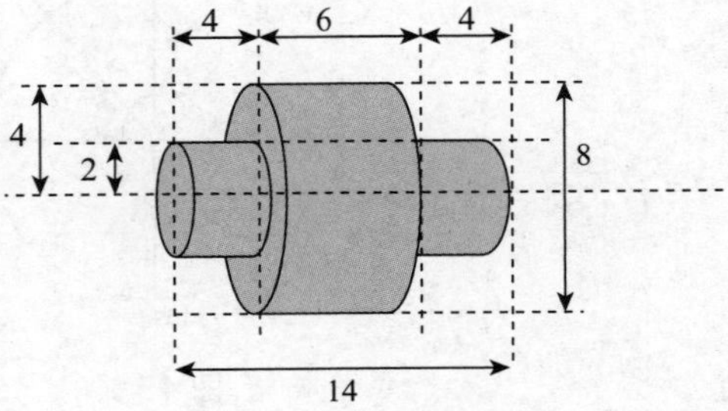 图4-5 锁插针的体积 $V=$ 两个小圆柱体的体积 $V_{小}+$ 一个大圆柱体的体积 $V_{大}$，即 $V = V_{小} \times 2 + V_{大} = \pi r_{小}^2 \times h_{小} \times 2 + \pi r_{大}^2 \times h_{大} = \pi \times 4 \times 4 \times 2 + \pi \times 16 \times 6 = 128\pi \approx 402$。
11	B	【题干】：从互联网上下载一首典型歌曲的文件大小为4兆字节。Lindy使用卫星网络在她的电脑下载歌曲，下载的网速为每秒256千字节，1兆字节=1,024千字节。 【问题】：在两小时内Lindy可以下载多少首歌曲？ 【解析】：根据题意，此题的关键是不同单位之间的转换。 两小时内可以下载的量为（以千字节计算）：$256 \times 60 \times 60 \times 2 = 1,843,200$千字节，转换为兆字节为 $\frac{1,843,200}{1,024} = 1,800$ 兆字节。一首典型歌曲的文件大小为4兆字节，共计为 $\frac{1,800}{4} = 450$ 首歌，选B。
12	C	【题干】：用韦恩图表示30个学生参加科学课。其中有人研究蝴蝶，有人研究蝗虫，有两种生物都研究，还有人一种生物也不研究。 【问题】：仅研究蝴蝶的学生占比是多少？ 【解析】：根据图中给出的信息，只研究蝴蝶的为9人，因此占整体的比例为 $\frac{9}{30} \times 100\% = 30\%$，选C。
13	B	【题干】：在一个沙滩上有一块长方形的游泳区域，边长分别为 x 和 y，面积是4,000。如图所示，用绳子将三个边围起来，海滩为其第四个边。现在需要用绳子将该区域分割为更小的三块长方形面积。 【问题】：分割成图中三块更小面积的区域，如果用 y 来表示，则需要绳子的总长度是多少？ 【解析】：根据题意设绳子的总长度为 L，可得，满足两个代数式： $$L = 4x + y$$ $$x \times y = 4,000$$ 将两个式子合并可得 $L = y + 4 \times \frac{4,000}{y} = y + \frac{16,000}{y}$，选B。

第四部分 数学必做100题

14	100	**【题干】**：一段长为80英寸的长形纸条片一端如图所示。其边缘有刻痕，如图中的加粗部分。现在将长形纸条片一条边每小段为4英寸的刻痕处去掉一个等边三角形。 **【问题】**：这段长为80英寸的长形纸条片新的刻痕边缘长度为多少？ **【解析】**：长形纸条长度为80英寸，在去掉三角形之前，尺寸如图所示，每小段长度为$3+1=4$英寸，所以一共有20段。现在去掉等边三角形之后，3英寸的部分不变，1英寸的部分增加了两个边，这两个边的长度即2英寸，所以每小段尺寸为$3+2=5$英寸。这样总的尺寸为$5\times20=100$英寸。
15	C	**【题干】**：1个四位登录码（0-9），要符合以下条件：1）首位数为素数；2）第二位数为奇数；3）第3位数能被5整除；4）第4位数为非素数奇数。 **【问题】**：这种类型的登录码有几种可能？ **【解析】**：根据题意可得，满足首位条件的数字有2，3，5，7，满足第二位条件的数字有1，3，5，7，9，满足第三位条件的数字有0和5，满足第四位条件的数字有1和9，所以有：$4\times5\times2\times2=80$种可能，选C。
16	2	**【题干】**：x^n是上式的最简形式。 **【问题】**：n的值是多少？ **【解析】**：根据题意可得$x^n=\dfrac{x^{\frac{1}{2}}\cdot x^{\frac{5}{6}}\cdot x^1}{x^{\frac{1}{3}}}$，$\dfrac{1}{2}+\dfrac{5}{6}+1-\dfrac{1}{3}=2$。
17	A	**【题干】**：1位汽车销售员的提成为每辆车价格的百分之k。 **【问题】**：当该汽车销售员卖了两辆车，每辆车卖$\$14,000$，下面哪个表达式表示他的提成量？ **【解析】**：根据题意可得，提成量$=14,000\times2\times k\%=\dfrac{14,000\times2\times k}{100}=280k$，选A。
18	B	**【题干】**：不同类型的灯泡如表所示，表中第一列为灯泡类型，第二列为价格，第三列为寿命。 **【问题】**：以美分/小时寿命来计算，紧凑型荧光灯（CFL）与发光二极管（LED）的价格比是多少？ **【解析】**：根据题意，首先需要进行美元与美分的换算。CFL的美分/小时价格为$\dfrac{1.50\times100}{10,000}=0.015$，LED的美分/小时价格为$\dfrac{10\times100}{25,000}=0.04$，比值为$\dfrac{0.015}{0.04}=\dfrac{3}{8}$，选B。
19	3,600	**【题干】**：如图所示，一个鱼缸底部铺了沙子，如果水面离缸顶的高度是3英寸。 **【问题】**：有多少立方英寸的水灌进了鱼缸？ **【解析】**：根据题意可得，该鱼缸是一个立方体，底面积为$30\times12=360$立方英寸。灌入的水的体积也是一个立方体。从图中可以看出鱼缸的高为15英寸，其中底部的沙子高为2英寸，水面里离缸顶的高度是3英寸，则灌入的水对应实际高度为$15-3-2=10$英寸。因此灌入的水体积为$360\times10=3,600$立方英寸。
20	D	**【题干】**：在学院建筑学书中有一张某建筑的图纸。在该书中，每半英寸代表实际建筑25英尺，如果书中某建筑的实际高度为350英尺。 **【问题】**：该建筑在书中图上的高度为多少？ **【解析】**：根据题意，图上尺寸与实际尺寸之间存在一定的比例关系，设该建筑在书中图上的高度为h，可得：$\dfrac{0.5}{25}=\dfrac{h}{350}$，解得$h=7$，选D。

Practice 4

 1. At the beginning of January, Tom deposits A dollars into a non-interest-bearing bank account. If Tom withdraws d dollars from the account every month and makes no additional deposits, how much money, in dollars, will be in the account after m months?

A. $A - md$ B. $(A - m)d$ C. $A - \dfrac{m}{d}$ D. $A - \dfrac{d}{m}$

 2. For a light that has an intensity of 60 candles at its source, the intensity in candles, S, of the light at a point d feet from the source is given by the formula $S = \dfrac{60k}{d^2}$, where k is a constant. If the intensity of the light is 30 candles at a distance of 2 feet from the source, what is the intensity of the light at a distance of 20 feet from the source?

A. $\dfrac{3}{10}$ candle   B. $\dfrac{1}{2}$ candle

C. $1\dfrac{1}{3}$ candles   D. 2 candles

 3. On the day of the performance of a certain play, each ticket that regularly sells for less than $10.00 is sold for half price plus $0.50, and each ticket that regularly sells for $10.00 or more is sold for half price plus $1.00. On the day of the performance, a person purchases a total of y tickets, of which x regularly sell for $9.00 each and the rest regularly sell for $12.00 each. What is the amount paid, in dollars, for the y tickets?

A. $7y - 2x$ B. $12x - 7y$
C. $7y + 4x$ D. $7y + 5x$

4. The figure (Fig. 4-15) represents a rectangular parking lot that is 30 meters by 40 meters and an attached semicircular driveway that has an outer radius of 20 meters and an inner radius of 10 meters. If the shaded region is not included, what is the area, in square meters, of the lot and driveway?

A. $1,350\pi$ B. $1,200 + 400\pi$
C. $1,200 + 300\pi$ D. $1,200 + 150\pi$

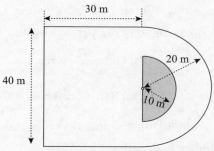

Note: Figure not drawn to scale

Fig. 4-15

5. The mean annual salary of an NBA player, S, can be estimated using the equation $S = 161,400(1.169)^t$, where S is measured in thousands of dollars, and t represents the number of years since 1980 for $0 \leqslant t \leqslant 20$. Which of the following statements is the best interpretation of 161,400 in the context of this problem?

A. The estimated mean annual salary, in dollars, of an NBA player in 1980

B. The estimated mean annual salary, in dollars, of an NBA player in 2000

C. The estimated yearly increase in the mean annual salary of an NBA player

D. The estimated yearly decrease in the mean annual salary of an NBA player

6. The owner of a spice store buys 3 pounds each of cinnamon, nutmeg, ginger, and cloves from distributor D (Table 4-7). She then sells all of the spices at $2.00 per ounce. What is her total profit, in dollars? (1 pound = 16 ounces)

A. $192   B. $282
C. $384   D. $486

Table 4-7   Spice Prices of Distributor D

Spice	Price Per Pound
Cinnamon	$8.00
Nutmeg	$9.00
Ginger	$7.00
Cloves	$10.00

7. The graph of $y = g(x)$ is shown in the figure (Fig. 4-16). If $g(x) = ax^2 + bx + c$ for constants a, b, and c, and if $abc \neq 0$, then which of the following must be true?

A. $ac > 1$ B. $c > 1$
C. $ac > 0$ D. $a > 0$

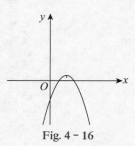

Fig. 4-16

8. Logan runs x miles per day. Which of the following represents the total distance he runs in a year, if he takes one day off per week, and a week off every 3 months?

A. $\left(52-\dfrac{12}{3}\right)\times 6x$ B. $\left(52-\dfrac{52}{3}\right)\times 6x$

C. $\left[365\times\left(\dfrac{6}{7}\right)-\dfrac{12}{3}\right]x$ D. $\left[365\times\left(\dfrac{6}{7}\right)-21\right]x$

9. A two-digit number from 10 to 99, inclusive, is chosen at random. What is the probability that this number is divisible by 5?

10. Find the value of x that is between 90° and 180° such that $\sin x° = \cos 30°$.

11. There are 75 more women than men enrolled in Linclen College. If there are n men enrolled, then, in terms of n, what percent of those enrolled are men?

A. $\dfrac{n}{2n+75}\%$ B. $\dfrac{n}{100\ (2n+75)}\%$

C. $\dfrac{100n}{n+75}\%$ D. $\dfrac{100n}{2n+75}\%$

12. When it is noon eastern standard time (EST) in New York City, it is 9:00 A. M. Pacific standard time (PST) in San Francisco. A plane took off from New York City at noon EST and arrived in San Francisco at 4:00 P. M, PST on the same day. If a second plane left San Francisco at noon PST and took exactly the same amount of time for the trip. What was the plane's arrival time (EST) in New York City?

A. 10:00 P. M, EST   B. 9:00 P. M, EST
C. 7:00 P. M, EST   D. 6:00 P. M, EST

Questions 13 – 14 refer to the following information.

A dietician working in a pediatrician's office is examining the diets of the children seen there. She has the record source about every child's consumption of fruits and vegetables each day for four weeks. She then averages the values for each child to find the grams of fruits and vegetables consumed daily by that child, rounded to the nearest quarter of a thousand. Each child's average consumption is marked as a dot on the graph (Fig. 4 – 17).

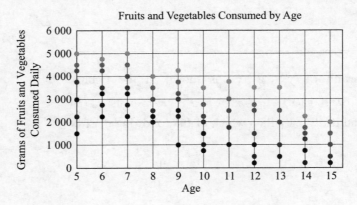

Fig. 4 – 17

13. The median consumption value for 8-year-olds is how much greater than the median consumption value for 13-year-olds?

   A. 2,750   B. 2,000   C. 1,250   D. 750

14. Which of the following is the closest to the mean consumption of fruits and vegetables, in grams, of children over 11 years of age?

   A. 1,500   B. 1,650   C. 2,350   D. 3,000

15. Recycling of certain metals has been a common practice dating back to preindustrial times. For example, there is evidence of scrap bronze and silver being collected and melted down for reuse in a number of European countries. Today, there are recycling companies and even curbside collection bins for recycling. As a general rule, recycling companies pay for metals by weight. Suppose a person brings in 3 pounds of copper and receives $8.64, and 24 ounces of nickel and receives $10.08. If another person brings in equal weights of copper and nickel, what fractional portion of the money would he receive from the copper?

16. If $g(x) = 2x^3 - 5x^2 + 4x + 6$, and P is the point on the graph of $g(x)$ that has an x-coordinate of 1, then what is the y-coordinate of the corresponding point on the graph of $g(x-3) + 4$?

Questions 17 - 18 refer to the following information.

Plants are capable of cross-pollinating with related but different plants. This creates a hybrid plant. Sometimes, a hybrid plant is superior to the two different plants from which it was derived. This is known as "hybrid vigor". Scientists can examine the DNA of a plant to see if it is a hybrid. This can be valuable information because if the plant appears superior, it would be beneficial to develop more of these hybrids. An agricultural scientist examines an orchard that has several types of apple trees and orange trees to see which ones are hybrids. Some of the information she collected is shown in the table (Table 4 - 8).

Table 4 - 8

	Apple Trees	Orange Trees	Total
Hybrid			402
Non-hybrid		118	
Total			628

17. Based on the data, if 45% of the apple trees are not hybrids, how many apple trees are hybrids?

   A. 50   B. 132   C. 226   D. 240

18. The scientist wants to study the orange trees to check for hybrid vigor. If she chooses one orange tree at random, what is the probability that it will be a hybrid?

A. $\frac{59}{194}$ B. $\frac{135}{314}$ C. $\frac{97}{157}$ D. $\frac{135}{194}$

19. During the typical Indy 500, the average pit stop is 15 seconds long and involves 6 crew members. Each of the 33 cars that race makes an average of 5 stops per race. If all cars start off at the same time and finish the race, using full crews and making the expected number of pit stops, what is the total number of active work hours put in by the crews during pit stops?

A. 4.125   B. 41.25   C. 206.25   D. 247.5

20. Mount Fuji in Japan was first climbed by a monk in 663 AD and subsequently became a Japanese religious site for hundreds of years. It is now a popular tourist site. When ascending the mountain, tourists drive part of the distance and climb the rest of the way. Suppose a tourist drove to an elevation of 2,390 meters and from that point climbed to the top of the mountain, and then descended back to the car taking the same route. If it took her a total of 7 hours to climb up and back down, and she climbed at an average rate of 264 vertical meters per hour going up and twice that going down, approximately how tall is Mount Fuji?

A. 1,386   B. 2,772   C. 3,622   D. 5,172

■ Practice 4 答案与解析

1	A	【题干】：在一月份开始，Tom 在不计利息的银行账户中存入 A 美元，每月从该账户中提现 d 美元，并且不再有新的资金存入。 【问题】：m 月后该账户中还有多少钱？ 【解析】：m 月后提取的总钱数为 $m \times d$，从总数 A 中扣除后，得到剩余的量为 $A - md$，选 A。
2	A	【题干】：一束光在光源处的强度为 60 烛光单位，距该光源 d 英尺地方的光强可以用公式 $S = \frac{60k}{d^2}$ 来表示，其中 k 是一个常数。 【问题】：若距离该光源 2 英尺远的地方的光强度是 30 烛光，则距离该光源 20 英尺远的地方的光强度是多少？ 【解析】：根据题意由"距离该光源 2 英尺远的地方的光强度是 30 烛光"，将 30 和 2 代入表达式：即 $30 = \frac{60k}{2^2}$，得 $k = 2$，再将 20 代入公式中，得到 $S = \frac{60 \times 2}{20^2} = \frac{3}{10}$，选 A。
3	A	【题干】：在某戏剧上演的当天，通常每张售价不到 $10 的票以正常票价的一半加上 $0.5 的价格出售，而通常每张售价在 $10 或者超过 $10 的票被以其正常票价的一半加上 $1 的价格出售。某人一共购买了 y 张票，其中 x 张票正常售价为 $9，其余票的正常售价为 $12。 【问题】：这个人购买这些票一共花了多少钱？ 【解析】：由题意：x 张正常售价为 $9 的票需要付 $x\left(\frac{9}{2} + 0.5\right)$；$y - x$ 张正常售价为 $12 的票需要付 $(y-x)\left(\frac{12}{2} + 1\right)$。所以总计付了 $5x + 7(y - x) = 7y - 2x$，选 A。

第四部分 数学必做100题

4	D	【题干】：图中表示一个面积为 30 米×40 米的长方形停车场，加上一个半圆形的汽车车道，其外部半径为 20 米，内部半径为 10 米。 【问题】：若不包括阴影区域，以平方米来计算，该停车场及车道的面积是多少？ 【解析】：根据题意，该面积＝长方形面积＋半圆面积（半径为 20）－阴影部分面积，根据条件可得：面积 $=30\times 40+\frac{1}{2}\pi(20^2-10^2)=1,200+150\pi$，选 D。
5	A	【题干】：NBA 球员平均年收入 S 可以用公式 $S=161,400(1.169)^t$ 表示，其中 S 以千来计，t 代表的是从 1980 年以来的年份，$0 \leqslant t \leqslant 20$。 【问题】：该式中 161,400 代表什么含义？ 【解析】：根据题意可得：在式子中，由于 t 可以取 0，当 $t=0$ 时，$S=161,400\times(1.169)^0=161,400\times 1=161,400$。表明是从 1980 年开始以来的第零年的 S 值，即 1980 年当年的 S 值。选 A。
6	B	【题干】：一位香料店店主从经销商 D 那购买了四种香料：肉桂、肉豆蔻、姜和丁香，每种香料各三磅，之后所有的香料都以 \$2.00/盎司出售。 【问题】：该店主的总利润是多少？ 【解析】：根据题意可得，该店主用于购买香料的费用是：$(8+9+7+10)\times 3=102$，出售香料的收入是：$2\times(4\times 3\times 16)=384$（注意有四种香料，每种三磅，并注意单位要转化为盎司）。所以总的利润＝出售香料的收入－购买香料的费用＝$384-102=282$，选 B。
7	C	【题干】：$y=g(x)$ 的图像如图所示。如果 $g(x)=ax^2+bx+c$，其中 a，b 和 c 都是常数，且 $abc\neq 0$。 【问题】：下列哪一项一定正确？ 【解析】：此图像为抛物线，a 的大小决定了抛物线的开口方向。从图中可以看出抛物线开口朝下，说明 $a<0$。此外，当 $x=0$ 时，$g(x)=g(0)=c$，代表的是抛物线在 y 轴上的截距。从图中可以看出图像与 y 轴的交点在 x 轴的下方，说明 $c<0$，因此 $a\times c>0$，选 C。
8	A	【题干】：Logan 每天跑 x 英里，其中每周休息一天，每三个月休息一周。 【问题】：下列哪个式子代表他一年里跑步的总路程？ 【解析】：根据题意可得，每三个月休息一周，一年共计休息 $\frac{12}{3}$ 周。已知一年一共有 52 周，因此实际跑步的周数为 $52-\frac{12}{3}$，其中每周跑步 $7-1=6$ 天，所以共计 $\left(52-\frac{12}{3}\right)\times 6$ 天。最后再乘以每天跑步的英里数，得到 A 项。
9	$\frac{1}{5}$ 或 0.2	【题干】：从包括 10 到 99 在内的两位数中，任意选择其中一个。 【问题】：能选择到可以整除 5 没有余数的两位数的可能性是多少？ 【解析】：根据题意，包括 10 到 99 在内的两位数，共计 90 个数，其中逢 10 的倍数（如 20，30 等），个位数为 5 的数（如 15，25）均满足该条件。共计 18 个。因此选择的可能性为 $\frac{18}{90}=\frac{1}{5}$ 或者 0.2。

10	120	【题干】：当 x 的值在 90°到 180°之间时，$\sin x° = \cos 30°$。 【问题】：x 的值为多少？ 【解析】：根据三角函数诱导公式可得 $\sin x° = \sin(180-x)° = \cos 30° = 0.5$，所以 $180 - x = 60$，解得 $x = 120$，如图 4-6。 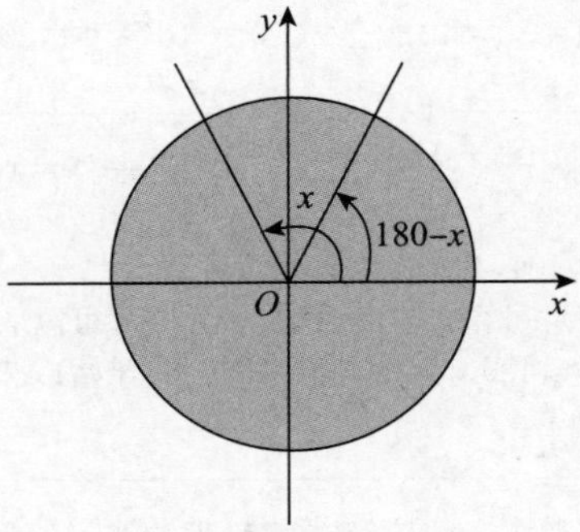 图 4-6
11	D	【题干】：在 Linden 学院注册的学生中女学生比男学生多 75，n 代表男学生的注册数量。 【问题】：以 n 来表示注册男学生的百分比是下面哪一个？ 【解析】：根据题意得，男学生为 n，因此女学生的数量为 $n+75$，总人数为 $n+n+75 = 2n+75$。可得男学生的百分比为 $\frac{n}{2n+75} \times 100\% = \frac{100n}{2n+75}\%$，选 D。
12	A	【题干】：纽约所在的东部时区是中午 12 时的时候，旧金山所在的太平洋时区是上午 9 时。一架飞机中午 12 时（东部时间）从纽约起飞，到达旧金山为当天下午 4 时（太平洋时间）。如果另一架飞机中午 12 时（太平洋时间）从旧金山起飞，以相同的飞行时间飞行。 【问题】：该架飞机何时（东部时间）到达纽约？ 【解析】：根据题意得，纽约所在的东部时区与旧金山所在的太平洋时区相差 12-9=3 小时的时差。一架飞机中午 12 时（东部时间）从纽约起飞，到达旧金山为当天下午 4 时（太平洋时间）即 16 时，换算成东部时间为 16+3=19 时，说明飞行的时间为 19-12=7 小时。另一架飞机中午 12 时（太平洋时间）从旧金山起飞，即 12+3=15 时（东部时间），加上飞行时间为 15+7=22 时，即晚上 10 点，选 A。
13	D	【题干】：儿科医师办公室有一位营养师检查孩子的饮食情况。她有四周内每个孩子每天摄取水果与蔬菜量的数据源。数据经过平均处理后可以得到每一位孩子水果与蔬菜日摄入量。数据近似处理以 250（1,000 的四分位）为一个梯级，如图所示。 【问题】：8 岁孩子摄入量的中位数值比 13 岁孩子的中位数值多多少？ 【解析】：根据题意，需要将两个年龄段孩子的数据先提取出来后再分析。 8 岁孩子的数据从小到大依次为 2,000，2,250，2,500，3,000，3,500 和 4,000，可得中位数为 $\frac{2,500+3,000}{2} = 2,750$。 13 岁孩子的数据从小到大依次为 500，1,000，2,000，2,500 和 3,500，可得中位数为 2,000，两者的差异为 2,750-2,000=750，选 D。
14	A	【题干】：参考第 13 题。 【问题】：下列哪一个数据更接近 11 岁以上孩子的水果与蔬菜日摄入量？ 【解析】：根题意可得，超过 11 岁的孩子包括了 12、13、14 和 15 岁，其中 12 岁共有七个值：250，500，1,000，1,500，2,500，2,750，3,500；13 岁共有五个值：500，1,000，2,000，2,500，3,500；14 岁共有六个值：250，750，1,250，1,500，1,750，2,250；15 岁共有五个值：250，500，1,000，1,500，2,000。最后将这 23 个数值取平均可得 1,500，选 A。

15	$\dfrac{3}{10}$	【题干】：自工业革命开始之前，回收材料就是一项常见的过程。比如有证据表明在很多欧洲国家从残留的青铜和银中收集和熔化掉金属以用于再利用。今天有很多的再生公司，甚至路边都设有用于回收的收集箱。一般来说，回收公司以金属的重量来付费，假设一个人带来3磅的铜，得到\$8.64，24盎司镍，得到\$10.08，如果另一个人带来了相同重量的铜和镍。 【问题】：这个人得到的钱中来自铜那一部分的收入所占比重是多少？ 【解析】：根据题意发现本题中重量单位不一致，所以需要统一单位。1磅=16盎司，因此铜的单价为 $\dfrac{8.64}{48}$ = \$ 0.18/盎司，镍的单价为 $\dfrac{10.08}{24}$ = \$ 0.42/盎司。现在两种金属的重量相等，可以设为1，因此铜那一部分的收入的比重是 $\dfrac{0.18\times 1}{0.18\times 1+0.42\times 1}=\dfrac{3}{10}$。 ("A fractional portion is part of a whole portion." 因此，在本题中由于有两种金属，所以考查的是其中某种金属所占的比重。)
16	11	【题干】：在函数 $g(x)$ 中，点 P 是 $g(x)$ 图像的某一点，且点 P 的横坐标值为1。 【问题】：当 $g(x)$ 变换为 $g(x-3)+4$ 时，点 P 对应的纵坐标值是多少？ 【解析】：根据题意可得，将1代入 $g(x)$，得到点 P 纵坐标值，即 $g(1)=7$。所以点 P 的坐标为（1，7）。当 $g(x)$ 变换为 $g(x-3)+4$ 时，图像向右移动了三个单位，向上移动了四个单位。所以点 P 坐标也相应地向右移动了三个单位，向上移动了四个单位。因此转换后的点 P 纵坐标值是 $7+4=11$。
17	B	【题干】：植物可以在不同植株之间异花受粉，这样就会产生杂交种。有时杂交种比这两个杂交的植物表现出更优的性状，这个现象称为"杂种优势"。科学家可以通过对DNA的测定来判断一棵植株是否为杂交种。这个信息非常有价值，如果该植株表现出更优的性状，就有价值来培育更多的杂交种。一个农学家检测了一个果园的几棵苹果树和桔子树来判断是否为杂交种。该农学家收集到的一些信息如表中所示。 【问题】：基于该表，如果45%的苹果树不是杂交种，则有多少棵苹果树是杂交种？ 【解析】：从表中给出的信息可以得到总计的非杂交种数量为 $628-402=226$，其中苹果树种非杂交种数量为 $226-118=108$。设苹果树一共有 A 棵，因此可得 $A\times 45\%=108$，解得 $A=240$，因此杂交的苹果树为 $240-108=132$，选B。
18	D	【题干】：背景信息参见17题。科学家希望研究桔子树来确定杂种优势，如果她随即挑选了一棵桔子树。 【问题】：问挑到的是杂交种桔子树的可能性为多少？ 【解析】：根据题意和17题的计算结果，苹果树一共有240棵，因此桔子树一共为 $628-240=388$ 棵，杂交种桔子树为 $388-118=270$ 棵，因此挑到的是杂交种桔子树的可能性为 $\dfrac{270}{388}=\dfrac{135}{194}$，选D。
19	A	【题干】：在一个典型的印第安500英里赛中，在每个停车加油点平均需要15秒时间，涉及6名机组人员。在比赛中33辆车每场比赛需要5次停车加油，如果所有的车同时开始且都完成了比赛，使用了所有的维修人员，实现了预期的停车加油环节。 【问题】：在停车加油环节总计花费维修人员多少工作时间？ 【解析】：根据题意可得，每个停车加油点6名机组人员，花费15秒，所以一共花费了 $15\times 6=90$ 秒，所以5次停车加油共需 $90\times 5=450$ 秒。33辆车共计 $450\times 33=14{,}850$ 秒 $=4.125$ 小时，选A。

| 20 | C | **【题干】**：公元 663 年，有僧侣攀登了日本富士山，数百年来富士山成为宗教场所。现在成为了一个著名的观光景点。爬山的时候，观光客首先开车到一定高度，然后再攀登剩余部分。假设观光客开车到海拔 2,390 米的地方，然后从该处爬到山顶。然后再沿原路返回达到停车的位置。如果爬山又下山花费了 7 小时的时间，她爬山的速度是 264 米/小时（垂直高度），下山的速度是上山的两倍。
【问题】：富士山大约有多高？
【解析】：根据题意可设爬山的时间为 x，下山的时间为 y，可得：$x+y=7$，且爬山的距离与下山的距离相等，即 $264 \times x = 528 \times y$，解得 $x \approx 4.67$，$y \approx 2.33$，所以爬山部分的高度为 $264 \times 4.67 \approx 1,232$ 米，富士山总计高度为 $1,232 + 2,390 = 3,622$ 米，选 C。 |

Practice 5

1. If $f(x) = x^2 - 11$, for what values of x is $f(x) < 25$?
   A. $-6 < x$
   B. $x < 6$
   C. $x \leqslant -6$ or $x \geqslant 6$
   D. $-6 < x < 6$

2. The table (Table 4-9) shows the value of an investment on January 1 of each year from 2005 to 2010. During which periods of time was the percent increase in the value of the investment the greatest?

   Table 4-9

Year	Value ($)
2005	150
2006	250
2007	450
2008	750
2009	1,200
2010	1,800

   A. 2005—2006
   B. 2006—2007
   C. 2008—2009
   D. 2009—2010

3. A certain theater has 100 balcony seats. For every $2 increase in the price of a balcony seat above $10, 5 fewer seats will be sold. If all the balcony seats are sold when the price of each seat is $10, which of the following could be the price of a balcony seat if the revenue from the sale of balcony seats is $1,360?
   A. $12   B. $14   C. $16   D. $17

4. Rectangular region *PQRS* shown in the figure (Fig. 4-18) is partitioned into ten identical smaller rectangular regions, each of which has width x. What is the perimeter of *PQRS* in terms of x?
   A. $15x$
   B. $25x$
   C. $30x$
   D. $50x$

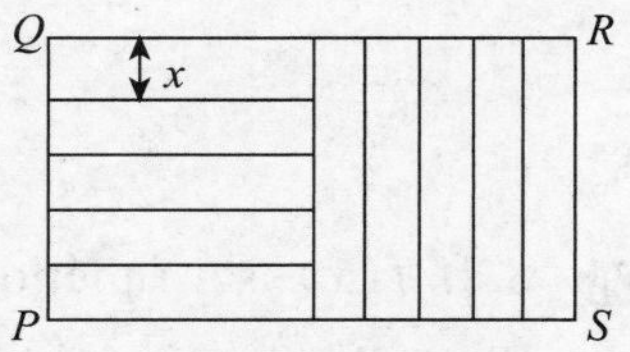

   Note: Figure not drawn to scale
   Fig. 4-18

5. A researcher wanted to know if there is an association between watching TV and

sleeping problems for the population of 15-year-olds in Germany. He obtained survey responses from a random sample of 2,200 15-year-olds Germans and found convincing evidence of a positive association between watching TV and sleeping problems.

Which of the following conclusions is well supported by the data?

A. There is a positive association between watching TV and sleeping problems for 15-year-olds in Europe

B. Using watching TV and sleeping problems as defined by the study, an increase in sleeping problems is caused by an increase in watching TV for 15-year-olds in Germany

C. Using watching TV and sleeping problems as defined by the study, an increase in sleeping problems is caused by an increase in watching TV for 15-year-olds in Europe

D. There is a positive association between watching TV and sleeping problems for 15-year-olds in Germany

6. The graph of line l in the xy-plane passes through the points $(2, 5)$ and $(4, 11)$. The graph of line m has a slope of -2 and an x-intercept of 2. If point (x, y) is the point of intersection of lines l and m, what is the value of y?

A. $\dfrac{3}{5}$ B. $\dfrac{4}{5}$ C. 1   D. 2

7. The figures (Fig. 4-19) show the graphs of the functions f and g: The function f is defined by $f(x) = 2x^3 + 5x^2 - x$. The function g is defined by $g(x) = f(x-h) - k$, where h and k are constants. What is the value of hk?

A. -2 B. -1 C. 0   D. 1

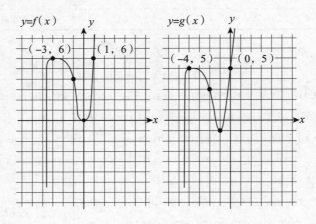

Fig. 4-19

8. If $f(x)$ is a function for which $f(2+k) = f(2-k)$, and $f(-3) = 0$, for which of the following is $f(x)$ also equal to zero?

A. $x = 7$ B. $x = 5$ C. $x = 3$ D. $x = -1$

9. If Aaron can do a job in 8 days, and Ben can do the same job in 12 days, how long does it take, in hours, for the two men, working together, to complete the same

job? (Round your answer to the nearest hour.)

10. If k is a rational number such that $k > 1$, which of the following could be the graph of the equation $y = ky + kx + x + 5$?

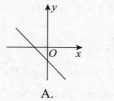 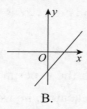 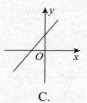

A.　　　　　　　B.　　　　　　　C.　　　　　　　D.

11. Phillip used four pieces of masking tape, each 6 inches long, to put up each of his posters. Phillip had a 300 feet roll of masking tape when he started. If no tape was wasted, which of the following represents the number of feet of masking tape that was left on the roll after he put up n posters? (12 inches = 1 foot)

A. $300 - 6n$　　B. $300 - 2n$　　C. $300 - n$　　D. $300 - \frac{1}{2}n$

12. Two airplanes at 25,000 feet above the ground are flying toward the same airport. Airplane A is flying due south toward the airport at 200 miles per hour, and Airplane B is flying due west toward the airport at 250 miles per hour. At 10:00 AM Airplane A is 700 miles from the airport, and Airplane B is 925 miles from the airport. How far are the two airplanes from each other at 10:30 AM?

A. 1,000 miles　B. 1,160 miles　　C. 1,400 miles　　D. 1,625 miles

Questions 13 through 15 refer to the following information.

John is conducting an experiment for his Economics class. Each morning on eight consecutive days, he sells homemade waffles in front of the school cafeteria. On the first day, he charges \$1 per waffle, and he raises the price by one dollar per day for the duration of the experiment. He records both the price and his net profits per day in the table (Table 4-10), but neglects to fill in his profits on days 2 and 7. To find the missing values for days 2 and 7, he writes a quadratic equation that accurately models the relationship between price and net profits.

Table 4-10

Price per waffle (in dollars)	Net profits
1	7
2	
3	15
4	16
5	15
6	12
7	
8	0

13. What is the sum of the net profits, in dollars, that John earned on days 2 and 7?
    A. 19   B. 20   C. 21   D. 2

14. John discovers that the quadratic equation that accurately models his results is of the form $N = -(p-a)^2 + c$, where p is the price, N is the net profits, and a and c are constants. What is the value of $a + c$?
    A. 20   B. 21   C. 22   D. 23

15. Suppose that John added a ninth day to the experiment, and raised the price of waffles to $9 per waffle. If the quadratic equation he wrote is accurate, what would John's net loss be, in dollars, on day 9?
    A. 9   B. 10   C. 11   D. 12

16. Anya shops at a thrift store only on sale days. On Mondays, members with a savings card receive a 25% discount on all items. On select Wedesdays, members receive a 30% discount on all items. The table (Table 4-11) shows the original costs of the items Anya bought one week that included the special Wedesday discount. A tax of 6.25% of the total purchase is applied to the total after all discounts. What is the total amount Anya spent at the thrift store this week, including tax?
    A. $17.41   B. $18.57   C. $19.49   D. $19.73

Table 4-11   Original Thrift Store Prices

Cost of items purchased on Monday	Cost of items purchased on Wednesday
$1.50	$0.98
$2.99	$1.99
$2.99	$2.55
$3.49	$4.98
$3.99	

17. A subway car on the New York City subway travels at an average speed of 17.4 miles per hour. Train cars on the Chicago L travel at an average speed that is 30% faster than that of the NYC subway. The DC Metro travels at an average speed that is 30% faster than that of the Chicago L. Marc rode the NYC subway from one stop to another and it took 6 minutes; Lizzie rode the Chicago L from one stop to another and it took 4.8 minutes; and Darius rode the DC Metro, which took 3.6 minutes between stops. How many miles did the person who traveled the shortest distance between stops travel? Round to the nearest tenth of a mile.

18. In an effort to decrease reliance on fossil fuels, some energy producers have started to utilize renewable resources. One such power plant uses solar panels to create solar energy during the day and then shifts to natural gas at night or when there is cloud cover. One particularly bright morning, the company increases the amount of its power typically

generated by solar panels by 60%. During a cloudy spell, it decreases the amount by 30%, and then when the sun comes back out, it increases the amount again by 75% before shutting the panels down for the night. What is the net percent increase of this company's reliance on solar panels during that day?

A. 75%   B. 96%   C. 105%   D. 165%

19. If s, t, u, and v are the coordinates of the indicated points on the number line (Fig. 4-20), which of the following is the greatest?

A. $|s-v|$ B. $|s-t|$ C. $|s+v|$ D. $|u+v|$

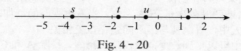

Fig. 4-20

20. In a survey of 80 students, 55 students stated that they play a varsity sport, and 35 stated that they are taking at least one AP level course. Which of the following statements must be true?

A. Less than half of the students who play a varsity sport are also taking at least one AP level course

B. At least one student who takes an AP level course does NOT play a varsity sport

C. The number of students who do not play a varsity sport is greater than the number of students who do not take at least one AP level course

D. Only 10 of these students are both playing a varsity sport and taking at least one AP level course

▌Practice 5 答案与解析

1	D	【题干与问题】：如果 $f(x)=x^2-11$，则 x 的值为多少时 $f(x)<25$？ 【解析】：根据题意：$f(x)<25 \Rightarrow x^2-11<25 \Rightarrow x^2<36$，解得 $-6<x<6$，选 D。
2	B	【题干】：表中列举了从 2005 年到 2010 年每年 1 月 1 号的投资额。 【问题】：哪一年的投资额增长率最大？ 【解析】：增长率＝实际增长值/变化起始值，可以根据此来计算： 2005 年—2006 年：$\frac{250-150}{150}=66.67\%$；2006 年—2007 年：$\frac{450-250}{250}=80\%$，2008 年—2009 年：$\frac{1,200-750}{750}=60\%$；2009 年—2010 年：$\frac{1,800-1,200}{1,200}=50\%$，投资额增长率最大为 2006 年—2007 年，选 B。
3	C	【题干】：某剧院有 100 个包厢座位，每一个包厢座位的单价在超过 \$10 时，每增加 \$2，剧院将少卖出 5 个座位，若包厢座位的单价为 \$10 时，所有的包厢座位都会被售出。 【问题】：若包厢座位的总收入为 \$1,360 时，其单价是下面哪一个？ 【解析】：根据题意可得，设票价增加 \$$x$，则此时少卖了 $\frac{5}{2}x$ 个座位。可得：$(10+x) \times \left(100-\frac{5}{2}x\right)=1,360$，解得 $x=16$ 或 34，选 C。

4	C	**【题干】**：图中所展示的长方形区域被分成10个宽相同且宽度均为 x 的小长方形区域。 **【问题】**：$PQRS$ 的周长用 x 来表示是多少？ **【解析】**：根据题意可得，大长方形的宽等于小长方形的长，为 $5x$，而大长方形的长等于小长方形长的两倍，为 $10x$，则 $PQRS$ 的周长为：$2\times(5x+10x)=30x$，选 C。
5	D	**【题干】**：有研究者想知道在德国15岁人群看电视与睡眠问题之间的关联，他从2,200位15岁德国人的随机样本中获得了调研回复，发现有令人信服的证据表明看电视与睡眠问题之间存在正相关。 **【问题】**：数据也可以支持以下哪个结论？ A. 欧洲15岁人中看电视与睡眠问题之间存在正相关 B. 从该研究的结果可得在德国15岁人中，当看电视时间增加时，会导致睡眠问题的增加 C. 从该研究的结果可得在欧洲15岁人中，当看电视时间增加时，会导致睡眠问题的增加 D. 在德国15岁人中，看电视与睡眠问题存在正相关 **【解析】**：从原题中可以得到数据范围仅限在德国，排除 A 和 C，并且看电视与睡眠问题存在正相关，但是不能推测出两者之间的因果关系，排除 B，选 D。
6	D	**【题干】**：在 xy 平面直角坐标系中，直线 l 通过 $(2,5)$ 和 $(4,11)$ 两点，直线 m 的斜率为 -2，在 x 轴上的截距为 2，点 (x,y) 是直线 l 与直线 m 相交的交点。 **【问题】**：y 的值是多少？ **【解析】**：根据题意首先可以根据"直线 l 通过 $(2,5)$ 和 $(4,11)$ 两点"这个条件得到直线 l 的斜率：$slop=\dfrac{11-5}{4-2}=3$，所以可以写出方程 $y=3x+b$。再根据直线 l 通过 $(2,5)$ 点，将 $x=2$，$y=5$ 代入 $y=3x+b$，可以求出 $b=-1$，所以直线 l 的方程为 $y=3x-1$。同理直线 m 的斜率已经给出，直线 m 在 x 轴上的截距为 2，说明直线 m 通过 $(2,0)$，可以得到直线 m 的方程为 $y=-2x+4$。 点 (x,y) 是直线 l 与直线 m 相交的交点，即满足方程组： $\begin{cases} y=3x-1 \\ y=-2x+4 \end{cases}$ 方程组的解 x 和 y 的值就是该点的坐标，通过消元解得 $x=1$，$y=2$，选 D。
7	B	**【题干】**：如图所示，给出了函数 $f(x)$ 和 $g(x)$ 的图像，其中 h 和 k 为常数。 **【问题】**：hk 的值是多少？ **【解析】**：由题意可得，$g(x)$ 的图像由 $f(x)$ 的图像在 xy 平面直角坐标系中经过上下左右移动得来。由 $f(x)$ 上的点 $(-3,6)$ 和 $g(x)$ 上的点 $(-4,5)$ 可得 $g(x)$ 的图像由 $f(x)$ 的图像向左移动了一个单位，向下移动了一个单位所得。因此 $g(x)$ 可以写成 $f(x+1)-1$ 的形式。对应原题的字母，可得 $h=-1$，$k=1$，所以 $hk=-1$，选 B。
8	A	**【题干】**：在函数 $f(x)$ 中满足条件 $f(2+k)=f(2-k)$，且 $f(-3)=0$。 **【问题】**：下面哪一个值使得 $f(x)$ 也为 0？ **【解析】**：根据题意，如果 $f(2+k)=f(-3)=0$，则 $2+k=-3$，解得 $k=-5$。那么也满足 $f(2-k)=f(2+k)=0$，即 $f(2-(-5))=f(7)=0$，选 A。
9	115	**【题干】**：Aaron 完成一项工作需要8天，Ben 完成同样的工作需要12天。 **【问题】**：当他们两人一起完成这份工作需要几小时（保留到整数）？ **【解析】**：根据题意可得，Aaron 完成一项工作需要8天，则他的工作效率为 $\dfrac{1}{8}$，同理，Ben 的工作效率为 $\dfrac{1}{12}$。因此两个人一起工作，工作效率为 $\dfrac{1}{8}+\dfrac{1}{12}=\dfrac{5}{24}$，因此总计需要 $\dfrac{1}{\frac{5}{24}}$ 即 $\dfrac{24}{5}$ 天完成。注意最后还要换算成小时，即 $\dfrac{24}{5}\times 24=115.2\approx 115$ 小时。

10	A	【题干】k 是一个有理数，且 $k>1$。 【问题】下面哪一个是函数 $y = ky + kx + x + 5$ 的图像？ 【解析】根据题意，将原函数化简为 $y = ax + b$ 的标准形式：$y = \dfrac{k+1}{1-k}x + \dfrac{5}{1-k}$，原题中给出 $k>1$，所以函数的斜率 $a = \dfrac{k+1}{1-k}<0$，函数在 y 轴上的截距 $b = \dfrac{5}{1-k}<0$，判断图像应该为 A。
11	B	【题干】Phillip 用四条胶条（每个胶条长 6 英寸），贴一张海报。开始的时候胶条的总长度为 300 英尺，贴海报的过程中没有胶条浪费。 【问题】下面哪一个式子表示当贴完 n 张海报时，胶条的剩余长度（英尺）？ 【解析】根据题意可得，贴一张海报需要的胶条数（以英尺来计算）为 $\dfrac{6\times 4}{12}$，n 张海报需要 $\dfrac{6\times 4}{12}\times n$，剩余的胶条长为 $300 - \dfrac{6\times 4}{12}\times n = 300 - 2n$，选 B。
12	A	【题干】两架飞机以离地面高度 25,000 英尺飞向同一个机场。飞机 A 从正南方向飞向机场，速度为 200 英里/小时，飞机 B 从正西方向飞向机场，速度为 250 英里/小时。上午 10 时，飞机 A 离机场 700 英里，飞机 B 离机场 925 英里。 【问题】上午 10:30 时，两架飞机互相之间的距离是多少？ 【解析】根据题意得，从 10:00 到 10:30，一共经历了 0.5 小时，在此期间飞机 A 飞行了 $200\times 0.5 = 100$ 英里，飞机 B 飞行了 $250\times 0.5 = 125$ 英里。此时飞机 A 离机场距离为 $700 - 100 = 600$ 英里，飞机 B 离机场距离为 $925 - 125 = 800$ 英里，正好形成一个直角三角形的两个直角边，而两架飞机之间的距离正好是这个直角三角形的斜边，根据勾股定理可得为 1,000 英里，选 A。
13	A	【题干】John 为他的经济学课程做了一个实验。连续八天早晨，他在学校的自助餐厅卖自制华夫饼。第一天，华夫饼每个卖 \$1，从第二天起价格每天增加 \$1，一直到实验结束。他将价格与净利润列在表中。但是第二天和第七天的记录有缺失。为了得到这些缺失的数据记录，他用一个二次方程式精确地模拟了价格与净利润之间的关系。 【问题】第二天和第七天 John 获得的净利润之和是多少？ 【解析】根据题意可得，价格与净利润之间的关系符合二次方程式，同时也发现在价格为 \$4 时，有最大的净利润 \$16，并且其他净利润都是围绕着 \$16 呈现一定的"对称"变化规律。可见价格为 \$2 时候的净利润应该与价格为 \$6 时候的相等，为 12；同理价格为 \$7 时候的净利润应该与价格为 \$1 时候的相等，为 7。因此第二天和第七天 John 获得的净利润是 $12+7=19$，选 A。
14	A	【题干】背景条件参见第 13 题。John 发现二次方程式 $N = -(p-a)^2 + c$ 可以精确地描述他的数据。该式子中 p 代表价格，N 代表净利润，a 和 c 都是常数。 【问题】$a + c$ 的值是多少？ 【解析】根据题意 $N = -(p-a)^2 + c$，说明该二次方程式的图像开口向下，因此当 $p = a$ 时有最大值 c，从原题的条件可以发现最大值是 16，所以 $c = 16$，对应 $p = 4$，因此 $a = 4$，计算得到 $a + c = 20$，选 A。
15	A	【题干】背景条件参见第 13 题。假设在实验中增加第九天，华夫饼的价格为 \$9，如果他写的二次方程式是正确的。 【问题】在第九天的时候 John 损失多少钱？ 【解析】根据题意结合第 14 题结果可得，原式 $N = -(p-4)^2 + 16$，把价格 $p = 9$ 代入该式得到 $N = -9$，说明净利润为负值，即损失 \$9，选 A。

16	D	【题干】：Anya 在一家二手货商店有折扣时去购物。周一的时候，会员用打折卡可以享受所有商品 25% 的折扣，在某些指定的周三，会员享受 30% 的折扣。表中给出的是 Anya 在一周内（包括一个特定的周三打折活动）所买商品的原始价格，在折扣完后所有商品需要征收 6.25% 的税。 【问题】：Anya 一共在这家二手货商店消费了多少钱（包括税）？ 【解析】：根据题意可得，周一 Anya 消费 \$14.96，打折后需要支付费用为 14.96×75% = \$11.22；周三 Anya 消费 \$10.5，打折后需要支付费用为 10.5×70% = \$7.35，共计消费为 11.22 + 7.35 = 18.57。加上税后，需要支付 18.57×(1+6.25%) = \$19.73，选 D。
17	1.7	【题干】：一列地铁列车在纽约地铁行驶的速度为 17.4 英里/小时，地铁列车在芝加哥的速度比纽约的快 30%，而在华盛顿特区地铁列车的速度又比芝加哥的快 30%。Marc 乘坐纽约地铁从一站到另一站用时 6 分钟；Lizzie 乘坐芝加哥地铁从一站到另一站用时 4.8 分钟；Darius 乘坐华盛顿特区地铁从一站到另一站用时 3.6 分钟。 【问题】：上述三人中哪个人的乘车距离最短？（保留到十分之一英里） 【解析】：根据题意可得，Marc 在纽约乘车距离为 $17.4 \times \frac{6}{60} = 1.74$ 英里，Lizzie 在芝加哥乘车距离为 $17.4 \times (1+30\%) \times \frac{4.8}{60} = 1.81$ 英里，Darius 在华盛顿特区乘车距离为 $17.4 \times (1+30\%) \times (1+30\%) \times \frac{3.6}{60} = 1.76$ 英里，因此 Marc 在纽约乘车距离最短，为 1.74 英里≈1.7 英里。
18	B	【题干】：为了减少对化石燃料的依赖性，一些能源生产者开始使用可再生资源。一家能源生产工厂白天使用太阳能板制备太阳能，晚上或者有云遮盖的时候就切换成天然气。在一个特别明亮的早晨，该公司通过使用太阳能板产能提高了 60%，而在有云遮盖的时候，产能下降了 30%，直到太阳重新出现的时候，产能提高了 75% 并且一直持续到晚上太阳能板收起来为止。 【问题】：在整个白天该公司的产能净增长的百分比是多少？ 【解析】：根据题意，由于该题没有给出具体的值，可以设一开始该公司的产能为 100，提高 60% 之后，产能为 100×(1+60%) = 160；下降了 30% 之后产能为 160×70% = 112；又增加了 75% 后产能为 112×(1+75%) = 196。所以产能净增长百分比为 $\frac{196-100}{100} \times 100\% = 96\%$，选 B。
19	A	【题干】：s, t, u, v 是数轴上对应的点。 【问题】：下面哪一个值最大？ 【解析】：根据题意可以从数轴上估读数字：$s = -3.8, t = -1.9, u = -0.5, v = 1.2$。因此 A 项：$|s-v| = |-3.8-1.2| = 5$；B 项：$|s-t| = |-3.8-(-1.9)| = 1.9$；C 项：$|s+v| = |-3.8+1.2| = 2.6$；D 项：$|u+v| = |-0.5+1.2| = 0.7$，所以 A 项最大，选 A。
20	D	【题干】：一项针对 80 名学生的调查表明，55 名学生说参加了一项大学体育运动项目，35 名同学说参加了至少一门 AP 课程。 【问题】：下面哪个表述一定正确？ A 项：参加大学体育运动项目的学生中不足一半的人同时也参加至少一门 AP 课程 B 项：至少有一名学生虽然参加了至少一门 AP 课程，但是没有参加大学体育运动项目 C 项：没有参加大学体育运动项目的人数要多于没有参加至少一门 AP 课程的人数 D 项：仅有 10 名学生既参加大学体育运动项目又至少参加一门 AP 课程 【解析】：根据题意，55 名学生说参加了一项大学体育运动项目，35 名同学说参加了至少一门 AP 课程，总计为 55+35 = 90，大于总数 80 人，这说明有 90-80 = 10 名学生是同时参加了两项的，D 项正确。至少参加一门 AP 课程的学生共计 35 名，可能也参加了大学体育运动项目，而 35 大于 55 的一半，因此 A 项错误。至少参加了一门 AP 课程人数为 35 人，没有参加大学体育运动项目的学生人数为 80-55 = 25 人，35+25 = 60<80，说明这两个部分没有重复，因此没有学生满足"虽然参加了至少一门 AP 课程，但是没有参加大学体育运动项目"，因此 B 项错误。没有参加大学体育运动项目的学生人数为 80-55 = 25 人，没有至少参加一门 AP 课程的学生人数为 80-35 = 45 人，25<45，因此 C 项错误，选 D。

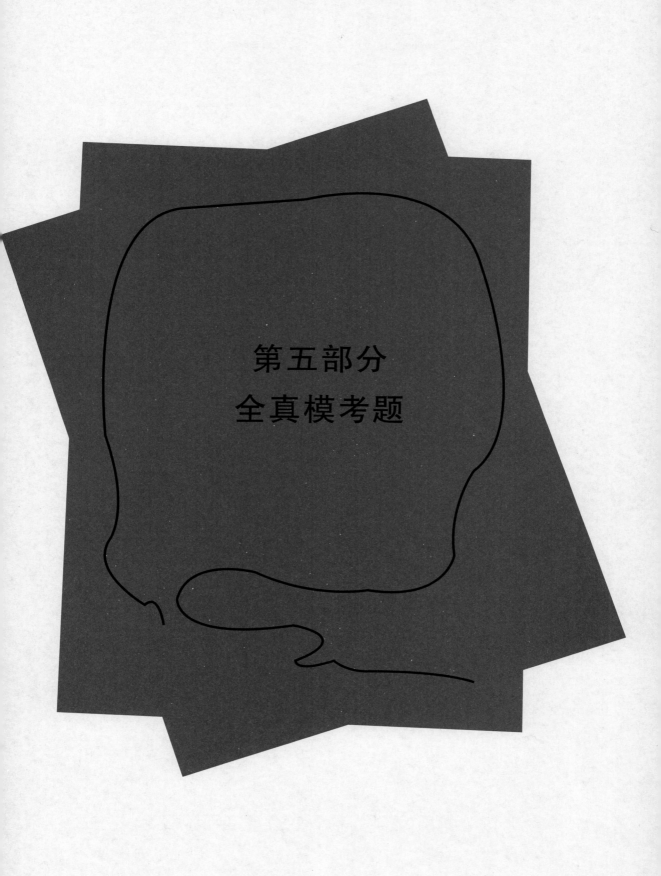
第五部分
全真模考题

Test 1

Section 3

Math Test—No Calculator
25 MINUTES, 20 QUESTIONS

Turn to Section 3 of your answer sheet to answer the questions in this section.

DIRECTIONS

For questions 1—15, solve each problem, choose the best answer from the choices provided, and fill in the corresponding circle on your answer sheet. For questions 16—20, solve the problem and enter your answer in the grid on the answer sheet. Please refer to the directions before question 16 on how to enter your answers in the grid. You may use any available space in your test booklet for scratch work.

NOTES

1. The use of a calculator is not permitted.
2. All variables and expressions used represent real numbers unless otherwise indicated.
3. Figures provided in this test are drawn to scale unless otherwise indicated.
4. All figures lie in a plane unless otherwise indicated.
5. Unless otherwise indicated, the domain of a given function f is the set of all real numbers x for which $f(x)$ is a real number.

REFERENCE

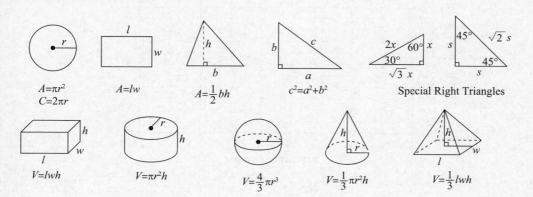

The number of degrees of arc in a circle is 360.
The number of radians of arc in a circle is 2π.
The sum of the measures in degrees of the angles of a triangle is 180.

Which of the following number lines represents the solution to the inequality $3x + 29 > 5 - x$?

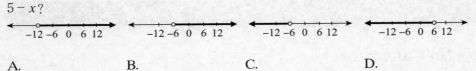

A.　　　　　　　B.　　　　　　　C.　　　　　　　D.

Marvin invested in a stock whose value had increased 5% by the end of 2012, decreased 2% by the end of 2013, and increased 10% by the end of 2014. What is the percent increase in the value of the stock from the end of 2012 to the end of 2014?

A. 7.80%　　　B. 8.19%　　　C. 12.50%　　　D. 13.00%

If $-\dfrac{3}{4} < 1 - 2t < -\dfrac{1}{4}$, what is one possible value of $8t - 4$?

A. 0　　　　　B. 1　　　　　C. 2　　　　　D. 3

⭐ 4

How many of the eight employees have an actual weekly salary that differs by more than $150 from the weekly salary predicted by the line (Fig. 5-1) of best fit?

A. 2　　　　　B. 3　　　　　C. 4　　　　　D. 6

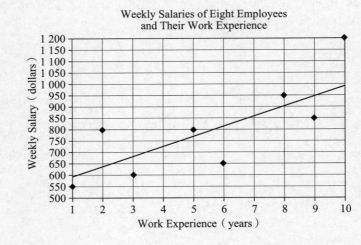

Fig. 5-1

If C is the temperature in degrees Celsius and F is the temperature in degrees Fahrenheit, then the relationship between temperatures on the two scales is expressed by the equation $9C = 5(F - 32)$. On a day when the temperature extremes recorded at a certain weather station differed by 45 degrees on the Fahrenheit scale, by how many degrees did the temperature extremes differ on the Celsius scale?

A. 13　　　　　　B. 25　　　　　　C. 45　　　　　　D. 81

The current population of a town is 10,000. If the population, P, increases by 3.5% every six months, which equation could be used to find the population after t years?

A. $P = 10,000(1.035)^{\frac{t}{2}}$　　　　　　B. $P = 10,000(0.965)^{2t}$

C. $P = 10,000(1.035)^{2t}$　　　　　　D. $P = 10,000(0.965)^{\frac{t}{2}}$

A candy store owner mixes candy that normally sells for \$15.00 per pound and candy that normally sells for \$22.50 per pound to make a 90-pound mixture to sell at \$18.00 per pound. To make sure that \$18.00 per pound is a fair price, how many pounds of the \$15.00 candy should the owner use?

A. 36　　　　　　　　　　B. 38
C. 42　　　　　　　　　　D. 54

Which of the following statements is true concerning the equation below?
$$3(5 - 2x) = 6(2 - x) + 3$$

A. The equation has no solutions

B. The equation has one positive solution

C. The equation has one negative solution

D. The equation has infinitely many solutions

A man drove his automobile d_1 kilometers at the rate of r_1 kilometers per hour an additional d_2 kilometers at the rate of r_2 kilometers per hour. In terms of d_1, d_2, r_1, r_2, what was his average speed, in kilometers per hour, for the entire trip?

A. $\dfrac{d_1 + d_2}{\dfrac{d_1}{r_1} + \dfrac{d_2}{r_2}}$　　B. $\dfrac{d_1}{r_1} + \dfrac{d_2}{r_2}$　　C. $\dfrac{\dfrac{d_1}{r_1} + \dfrac{d_2}{r_2}}{d_1 + d_2}$　　D. $\dfrac{r_1 + r_2}{d_1 + d_2}$

If $y = 3^x$, which of the following expressions is equivalent to $9^x - 3^{x+2}$ for all positive integer values of x?

A. $3y - 3$
B. $y^2 - y$
C. $y^2 - 3y$
D. $y^2 - 9y$

The figure (Fig. 5-2) shows the graph of the linear function $y = f(x)$. If the slope of the line is -2 and $f(3) = 4$, what is the value of b?

A. 8   B. 9   C. 10   D. 11

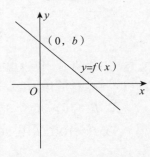

Note: Figure not drawn to scale

Fig. 5-2

If a factory can manufacture b computer screens in n days at a cost of c dollars per screen, then which of the following represents the total cost, in dollars, of the computer screens that can be manufactured, at that rate, in m days?

A. $\dfrac{bcm}{n}$ B. $\dfrac{bmn}{c}$ C. $\dfrac{mc}{bn}$ D. $\dfrac{bc}{mn}$

Given the table (Table 5-1) of values for functions g and h, for what value of x must $g(h(x)) = 6$?

A. 2   B. 5   C. 6   D. 12

Table 5-1

x	$g(x)$	$h(x)$
1	2	-9
2	4	-6
3	6	-3
4	8	0
5	10	3
6	12	6
7	14	9
8	16	12
9	18	15

14. If $0 < x < \frac{\pi}{2}$ and $\frac{\cos x}{1-\sin^2 x} = \frac{3}{2}$, what is the value of $\cos x$?

A. $\frac{1}{9}$ B. $\frac{1}{3}$ C. $\frac{4}{9}$ D. $\frac{2}{3}$

15. In the diagram (Fig. 5 – 3) of circle C, chord $\overline{PQ}$ intersects chord $\overline{RS}$ at T. If $PQ = 4x + 6$, $TQ = 5$, $RS = 6x + 8$, and $TS = 3$, what is the value of x?

A. 3 B. 5
C. 10 D. 21

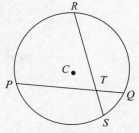

Note: Figure not drawn to scale
Fig. 5 – 3

DIRECTIONS

For questions 16—20, solve the problem and enter your answer in the grid, as described below, on the answer sheet.

1. Although not required, it is suggested that you write your answer in the boxes at the top of the columns to help you fill in the circles accurately. You will receive credit only if the circles are filled in correctly.

2. Mark no more than one circle in any column.

3. No question has a negative answer.

4. Some problems may have more than one correct answer. In such cases, grid only one answer.

5. Mixed numbers such as $3\frac{1}{2}$ must be gridded as 3.5 or 7/2. (If is entered into the grid, it will be interpreted as $\frac{31}{2}$, not as $3\frac{1}{2}$.)

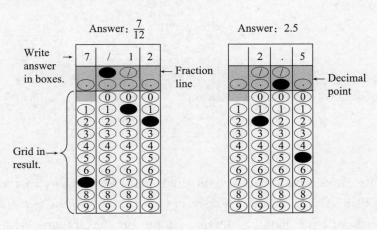

6. Decimal answers: If you obtain a decimal answer with more digits than the grid can accommodate, it may be either rounded or truncated, but it must fill the entire grid.

Acceptable ways to grid $\frac{2}{3}$ are:

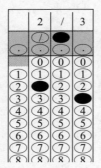  

Answer: 201-either position is correct

 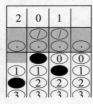 NOTE: You may start your answers in any column, space permitting. Columns you don't need to use should be left blank.

In a writer's workshop, there are half as many men as women. If there are 24 total men and women in the writer's workshop, how many women are there?

If (x, y) is a solution to the system of equations below and y does not equal 0, what is the value of $(y - x)^3$?

$$x^2 - y^2 = y$$
$$x = -3y$$

What is one possible solution to the equation?

$$\frac{6}{x+1} - \frac{3}{x-1} = \frac{1}{4}?$$

A bank offers its business customers two different plans for monthly charges on checking accounts. With Plan A, the account holder pays a fee of $0.15 per check processed during the month with no monthly service charge on the account. With Plan B, the account holder pays a $10.00 monthly service charge with a fee of $0.05 per check processed during the month. Find the break-even point for the two plans. That is, find the number of checks processed per month for which the costs of the two plans are equal.

The table (Table 5-2) shows the one-way driving distance, in miles, between four cities: P, Q, R, and S. For example, the distance between P and Q is 144 miles. If the round trip between S and Q is 16 miles further than the round trip between S and R, and the round trip between S and R is 24 miles less than the round trip between S and P, what is the value of X?

Table 5-2

	P	Q	R	S
P	0	144	171	186
Q	144	0	162	X
R	171	162	0	Y
S	186	X	Y	0

STOP

If you complete this section before the end of your allotted time, check your work on this section only. Do NOT use the time to work on another section.

Section 4

Math Test—Calculator

55 MINUTES, 38 QUESTIONS

Turn to Section 4 of your answer sheet to answer the questions in this section.

DIRECTIONS

For questions 1—30, solve each problem, choose the best answer from the choices provided, and fill in the corresponding circle on your answer sheet. For questions 31—38, solve the problem and enter your answer in the grid on the answer sheet. Please refer to the directions before question 31 on how to enter your answers in the grid. You may use any available space in your test booklet for scratch work.

NOTES

1. The use of a calculator is permitted.
2. All variables and expressions used represent real numbers unless otherwise indicated.
3. Figures provided in this test are drawn to scale unless otherwise indicated.
4. All figures lie in a plane unless otherwise indicated.
5. Unless otherwise indicated, the domain of a given function f is the set of all real numbers x for which $f(x)$ is a real number.

REFERENCE

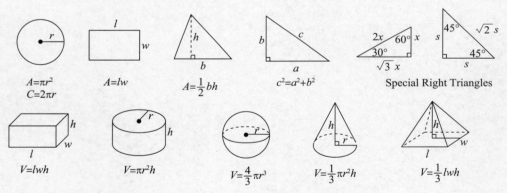

The number of degrees of arc in a circle is 360.
The number of radians of arc in a circle is 2π.
The sum of the measures in degrees of the angles of a triangle is 180.

CONTINUE

How many values of x satisfy the equation $x^2 - 8x = -16$?

A. none　　　　　B. 1　　　　　C. 2　　　　　D. more than 2

The tenth number of the sequence 50, 44.5, 39, 33.5, ⋯ is

A. -4　　　　　B. 0.5　　　　C. 1　　　　　D. 1.5

If $a < -1$, which of the following could be the graph of $y - 1 = \dfrac{a}{a+1}x$?

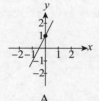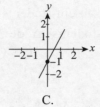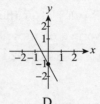

　　A.　　　　　　　B.　　　　　　　C.　　　　　　　D.

If the average of 5 positive integers is 70, what is the largest possible value of their median?

A. 70　　　　　B. 114　　　　C. 116　　　　D. 346

Questions 5 and 6 are based on the graph (Fig. 5 – 4).

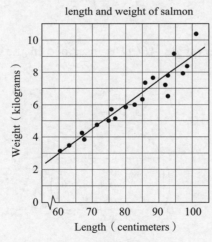

Fig. 5 – 4

The scatterplot above shows the length and weight of a group of 20 salmon and the line of best fit for the data. According to this line of best fit, which of the following best approximates the weight, in kilograms, of a salmon that is 95 centimeters long?

A. 7.6　　　　　B. 7.8　　　　C. 8.3　　　　D. 8.8

Which of the following equations best describes the relationship between w, the weight in kilograms of each salmon, and l, its length in centimeters?

A. $w = \dfrac{3}{20}l + 2$ 　　　　　　B. $w = \dfrac{20}{3}l + 2$

C. $w = \dfrac{3}{40}l - 6$ 　　　　　　D. $w = \dfrac{3}{20}l - 6$

There are 25 students in Mrs. Li's first period algebra class. On Monday, 5 students were absent and the other 20 students took a test. The average grade for those students was 86. The next day after the 5 absent students took the test, the class average was 88. What was the average of those 5 students' grades?

A. 90　　　　　　B. 92　　　　　　C. 94　　　　　　D. 96

If for all real numbers x, $g(3-x) = x^2 + x + 1$, what is the value of $g(7)$?

A. 13　　　　　　B. 21　　　　　　C. 57　　　　　　D. 111

If $g(x+1) = x^2 + 2x + 4$ for all values of x, which of the following is equal to $g(x)$?

A. $x^2 + 4$ 　　　　　　B. $x^2 + 3$

C. $(x-1)^2 + 4$ 　　　　　　D. $(x-1)^2 + 3$

In the figure (Fig. 5-5), rectangle $ABCD$ is inscribed in the circle with center O. What is the area of the circle?

A. 26π 　　　　　　B. 121π

C. 144π 　　　　　　D. 169π

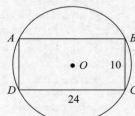

Note: Figure not drawn to scale

Fig. 5-5

A sphere and a cone have equal volumes. If the radius of the cone is twice the radius of the sphere, what is the ratio of the height of the cone to its radius?

A. 0.5 : 1　　　　B. 1 : 1　　　　C. 2 : 1　　　　D. π : 1

Questions 12 to 14 are based on the graph (Fig. 5-6).

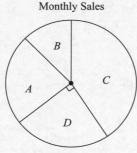

Fig. 5-6

The pie graph above represents the monthly ad sales for four sales people—Maria, Eli, Georgia, and Zoe—at a social media website. For the month, Maria's sales accounted for 25% of the total, Eli had $3,000 in sales, Georgia had $5,000 in sales, and Zoe had $10,000 in sales.

Which sector represents Georgia's sales for the month?
A. Sector A   B. Sector B   C. Sector C   D. Sector D

What is the sum of the monthly sales for all four sales people?
A. $22,500   B. $24,000
C. $25,000   D. $27,000

If Eli and Georgia both earn 10% commission on their sales, and Maria and Zoe both earn 15% commission on their sales, how much more did Maria earn in monthly commissions than Georgia?
A. $300   B. $360   C. $375   D. $400

Which of the following functions, when graphed in the xy-plane, will intersect the x-axis exactly 3 times?
A. $f(x) = (x^2+1)(x^2+1)$ B. $f(x) = (x^2-1)(x^2+1)$
C. $f(x) = x^2(x^2-1)$ D. $f(x) = x^2(x^2+1)$

A website received 2,100 visitors in July from both subscribers and non-subscribers. If the ratio of subscribers to non-subscribers among this group was 2 : 5, how many more non-subscribers visited the site in July than subscribers?
A. 126   B. 630   C. 900   D. 1,260

17

For what value of k will the graphs of $3x+4y+5=0$ and $kx+6y+7=0$ NOT intersect?

A. -8 B. 4.5 C. 5 D. 8

18

If $\dfrac{x^2+1}{2}+\dfrac{x}{p}=1$, which of the following expressions gives both possible values of x, in terms of p?

A. $\dfrac{-p\pm\sqrt{p^2-8p}}{2p}$ B. $\dfrac{-p\pm\sqrt{p^2-4p}}{2p}$

C. $\dfrac{-2\pm\sqrt{4+4p^2}}{2p}$ D. $\dfrac{-2\pm\sqrt{4-4p^2}}{2p}$

19

Class A: 68, 79, 88, 91, 97, 98, 99
Class B: 85, 85, 85, 88, 88, 90, 90

The lists above indicate the tests scores, in increasing order, for two of Mr. Pearlman's classes, each of which has 6 students. Which of the following correctly compares the standard deviation of the scores for each class?

A. The standard deviation of the scores in Class A is smaller
B. The standard deviation of the scores in Class B is smaller
C. The standard deviations of the scores in Class A and Class B are equal
D. The relationship cannot be determined from the information given

20

If $f(x)=x+2$ and $f(g(1))=6$, which of the following could be $g(x)$?

A. $g(x)=3x$ B. $g(x)=x+3$
C. $g(x)=x-3$ D. $g(x)=2x+1$

21

A cylindrical tube with negligible thickness is placed into a rectangular box that is 3 inches by 4 inches by 8 inches, as shown in the accompanying diagram (Fig. 5 – 7). If the tube fits exactly into the box diagonally from the bottom left corner to the top right back corner, what is the best approximation of the number of inches in the length of the tube?

A. 3.9 B. 5.5 C. 7.8 D. 9.4

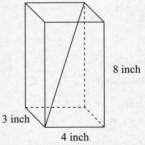

Note: Figure not drawn to scale

Fig. 5 – 7

22

Let the function f be defined by $f(x)=2-|x-4|$ for all real values of x. What is the greatest value of f?

A. -2 B. 2 C. 4 D. 6

The Horizon Resort charges $150 per night for a single room, and a one-time valet parking fee of $35. There is a 6.5% state tax on the room charges, but no tax on the valet parking fee. Which of the following equations represents the total charges in dollars, c, for a single room, valet parking, and taxes, for a stay of n nights at The Horizon Resort?

A. $c = (150 + 0.065n) + 35$ B. $c = 1.065 \times 150n + 35$
C. $c = 1.065(150n + 35)$ D. $c = 1.065(150 + 35)n$

If y varies directly as x and $y = 12$ when $x = c$, what is y in terms of c when $x = 8$?

A. $\dfrac{2c}{3}$ B. $\dfrac{3}{2c}$ C. $20c$ D. $\dfrac{96}{c}$

Questions 25 to 27 refer to the following information (Table 5-3).

Table 5-3   TALENT SHOW TICKETS

	Adult	Child	Senior	Student
Tickets Sold	84	40	16	110
Total Income	$630	$200	$96	$495

According to the table, how much is the price of one senior ticket?

A. $4.00   B. $6.00
C. $12.00   D. $16.00

How much more is the cost of one adult ticket than the cost of one student ticket?

A. $0.50   B. $1.50
C. $2.50   D. $3.00

Which is closest to the average (arithmetic mean) price of the all tickets sold?

A. $5.54   B. $5.59
C. $5.68   D. $5.72

Which of the following is NOT equal to $i^6 - i^2$?

A. $i^5 - i$ B. i^4
C. $2i^3 + 2i$ D. $1 + i^6$

29

If $\sin y = \dfrac{a}{b}$ and $0 < y < \dfrac{\pi}{2}$, which of the following is equal to $\sin\left(\dfrac{\pi}{2} - y\right)$?

A. $\dfrac{\sqrt{a^2 - b^2}}{a}$ B. $\dfrac{\sqrt{b^2 - a^2}}{a}$

C. $\dfrac{\sqrt{a^2 - b^2}}{b}$ D. $\dfrac{\sqrt{b^2 - a^2}}{b}$

30

If $f(x) = (x^2)^{-2b}$ and $f(3) = 3$, what is the value of b?

A. $-\dfrac{1}{2}$ B. $-\dfrac{1}{4}$ C. $\dfrac{1}{4}$ D. $\dfrac{1}{2}$

DIRECTIONS

For questions 31—38, solve the problem and enter your answer in the grid, as described below, on the answer sheet.

1. Although not required, it is suggested that you write your answer in the boxes at the top of the columns to help you fill in the circles accurately. You will receive credit only if the circles are filled in correctly.
2. Mark no more than one circle in any column.
3. No question has a negative answer.
4. Some problems may have more than one correct answer. In such cases, grid only one answer.
5. Mixed numbers such as $3\dfrac{1}{2}$ must be gridded as 3.5 or 7/2. (If ┌3│1│/│2┐ is entered into the grid, it will be interpreted as $\dfrac{31}{2}$, not $3\dfrac{1}{2}$.)
6. Decimal answers: If you obtain a decimal answer with more digits than the grid can accommodate, it may be either rounded or truncated, but it must fill the entire grid.

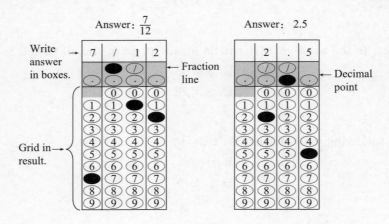

Acceptable ways to grid $\frac{2}{3}$ are:

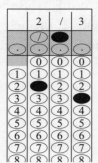  

Answer: 201–either position is correct

NOTE: You may start your answers in any column, space permitting. Columns you don't need to use should be left blank.

CONTINUE

The number of shells in Fred's collection is 80% of the number in Phil's collection. If Phil has 80 more shells than Fred, how many do they have altogether?

If $f(x) = 3 + \frac{5}{x}$, what number CANNOT be a value of $f(x)$?

If $\frac{x}{x+1} + \frac{1}{x-1} = \frac{25}{24}$ and $x > 0$, what is the value of x?

If the lines $y = -4x - 3$ and $y = -3x - b$ intersect at the point $(-1, c)$, what is the value of b?

A farmer purchased several animals from a neighboring farmer: 6 animals costing $100 each, 10 animals costing $200 each, and k animals costing $400 each, where k is a positive odd integer. If the median price for all the animals was $200, what is the greatest possible value of k?

Each step of a staircase is 0.25 meter wide and 0.20 meter high, as shown in the figure (Fig. 5-8). All angles shown in the figure are right angles. If the height of the staircase is 3.6 meters and the landing at the top of the staircase is 1 meter wide, how long, in meters, is AB?

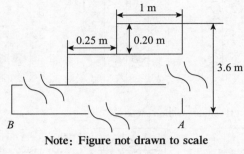

Note: Figure not drawn to scale

Fig. 5-8

Questions 37 and 38 refer to the following information.

An Internet service provider offers three different plans for residential users. Plan A charges users $500 for the first year of service, and $80 per month thereafter. Plan B charges users $68 per month. Plan C is a "high speed" plan that offers 200% higher speeds for $92 per month.

Isabelle has been using Plan A for over a year. She recently reviewed her plan and realized that if she had been using Plan B for same amount of time, she would have saved $104 for Internet service over the entire period. At the time of her review, how many months had Isabelle been on Plan A?

Isabelle is now considering switching to either Plan B or Plan C for her home business, but she calculates that having the "high speed" plan will save her only approximately 45 minutes of work each month. At what minimum hourly rate, in dollars per hour, would she have to value her work for Plan C to be more worth than Plan B? (Round your answer to the nearest whole number)

STOP

If you complete this section before the end of your allotted time, check your work on this section only. Do NOT use the time to work on another section.

第五部分 全真模考题

■ Test 1 答案与解析

		Section 3
1	B	【解析】：根据题意，解不等式后得到 $x > -6$，符合该条件的数轴图为 B。
2	A	【题干】：Marvin 投资了一只股票，到了 2012 年底股票的价值增加了 5%，但是到了 2013 年底下降了 2%，然后到了 2014 年底又增加了 10%。 【问题】：2012 年到 2014 年该只股票的价值增加了多少？ 【解析】：根据题意，假设以 100 为股票的初始值，到了 2012 年底为 $100 \times (1 + 5\%) = 105$，到了 2013 年底为 $105 \times 98\% = 102.9$，到了 2014 年底为 $102.9 \times 110\% = 113.19$。因此从 2012 年到 2014 年增加的百分比为 $\frac{113.19 - 105}{105} \times 100\% = 7.80\%$，选 A。
3	C	【题干与问题】：如果 $-\frac{3}{4} < 1 - 2t < -\frac{1}{4}$，$8t - 4$ 的一个可能值是多少？ 【解析】：根据题意，将不等式两边同时乘以 -4 可得 $1 < -4 + 8t < 3$。因此满足该不等式的只有 2，选 C。
4	B	【题干】：图中给出的是八位工人的工作经历（x 轴）与周薪（y 轴）的关系。 【问题】：有多少位工人的实际周薪与通过最适拟合线预测周薪的差值是大于 \$150 的？ 【解析】：根据题意，图中对应的纵坐标每一个格为 50，图中的点代表的是八位工人的实际周薪，要想找到比预测周薪的差值大于 \$150 的人数，只要找到这八个点中与其对应的最适拟合线之间的纵向距离超过三个格子的点即可。从图上看，满足条件的有三个点，因此选 B。
5	B	【题干】：C 代表摄氏温度，F 代表华氏温度，两种温度之间的关系用公式表示为：$9C = 5(F - 32)$。一天某气象站记录下华氏温度的最大温度值和最小温度值的差值为 45 度。 【问题】：那么摄氏温度的最大温度值和最小温度值的差值是多少？ 【解析】：根据题意，可以将温度的差值代入到公式中去：由 $C = \frac{5}{9}(F - 32)$，可得 $\Delta C = \frac{5}{9} \Delta F = \frac{5}{9} \times 45 = 25$，选 B。
6	C	【题干】：一个镇现在的人口为 1 万人，如果每六个月增加 3.5%。 【问题】：下面哪个式子代表 t 年之后的人口数？ 【解析】：根据题意：每六个月的增长系数为 $1 + 3.5\% = 1 + 0.035 = 1.035$，因此，人口数量＝起始人口数×(增长系数)^{时间周期}。以六个月为一个时间段，t 年一共有 $2t$ 个这样的时间段，因此符合条件的为 C 项，选 C。
7	D	【题干】：一家糖果店主将两种糖果混合在一起，其中一种糖果的价格为每磅 \$15.00，另一种的价格为每磅 \$22.50。混合后的糖果为 90 磅，价格为每磅 \$18.00。 【问题】：为了使每磅 \$18.00 是一个公平合理的价格，需要多少磅每磅价格为 \$15.00 的糖？ 【解析】：根据题意，设价格为每磅 \$15.00 的糖有 x 磅，价格为每磅 \$22.50 的糖有 y 磅，可以得到一个方程组： $$x + y = 90$$ $$15x + 22.5y = 18 \times 90$$ 解方程组可得 $x = 54$，选 D。

8	D	【题干与问题】：对下面式子的描述，哪一个是正确的？ 【解析】：根据题意： 该式子的左侧为 $3(5-2x)=15-6x$，右侧为 $6(2-x)+3=12-6x+3=15-6x$，可见原式两边恒等，因此每一个实数都可以满足上面的条件，选 D。
9	A	【题干】：一个人以 r_1 公里/小时的速度开车行驶了 d_1 公里。接着又以 r_2 公里/小时的速度行驶了 d_2 公里。 【问题】：他整个旅程的平均速度是每小时多少公里（用 d_1，d_2，r_1，r_2 表示）？ 【解析】：平均速度＝总路程/总时间。根据题意可得，这个人开车行驶的总路程为 d_1+d_2，开车所用的总时间为 $\frac{d_1}{r_1}+\frac{d_2}{r_2}$，所以整个旅程的平均速度为 $\frac{d_1+d_2}{\frac{d_1}{r_1}+\frac{d_2}{r_2}}$，选 A。
10	D	【题干】：已知 $y=3^x$。 【问题】：对于所有正整数而言，下列哪一个式子与 9^x-3^{x+2} 相等？ 【解析】：根据题意：$9^x-3^{x+2}=(3^2)^x-3^x\times 3^2=(3^x)^2-9\times(3^x)=y^2-9y$，选 D。
11	C	【题干】：图中显示的是线性函数 $f(x)$ 的图像，如果直线的斜率为 -2，且 $f(3)=4$。 【问题】：b 的值为多少？ 【解析】：根据题意，线性函数的标准式为：$y=mx+b$，其中 m 代表斜率，在本题中为 -2，再将 $x=3$，$y=4$ 代入该线性函数，得到 $b=10$，选 C。
12	A	【题干】：一家工厂在 n 天内生产了 b 台电脑显示器，每台成本为 c 美元。 【问题】：下面哪一个式子代表的是以此生产率生产 m 天的电脑显示器的总成本？ 【解析】：根据题意，每天的生产量为：$\frac{b}{n}$，再乘以单台的成本 c，再乘以 m 天，可得 $\frac{bcm}{n}$，选 A。
13	B	【题干】：表中给出了 $g(x)$ 和 $h(x)$ 的一系列的值。 【问题】：当满足 $g(h(x))=6$ 时，x 的值为多少？ 【解析】：根据题意，从表中可以得到当 $g(x)=6$ 时，x 取值为 3 时，再看 $h(x)=3$，x 取值为 5，选 B。
14	D	【题干】：如果角 x 大于 0 度，小于 90 度，且 $\frac{\cos x}{1-\sin^2 x}=\frac{3}{2}$。 【问题】：$\cos x$ 的值为多少？ 【解析】：根据题意可得：$\cos^2 x=1-\sin^2 x$，所以 $\frac{\cos x}{1-\sin^2 x}=\frac{\cos x}{\cos^2 x}=\frac{1}{\cos x}=\frac{3}{2}$，因此 $\cos x=\frac{2}{3}$，选 D。
15	B	【题干】：在圆 C 中，弦 PQ 与弦 RS 相交与点 T，$PQ=4x+6$，$TQ=5$，$RS=6x+8$，$TS=3$。 【问题】：x 的值为多少？ 【解析】：根据题意以及相交弦定理可得 $PT\times TQ=RT\times TS$。题中给出条件 $PT=PQ-TQ=4x+6-5=4x+1$，$RT=RS-TS=6x+8-3=6x+5$，代入上式得 $(4x+1)\times 5=(6x+5)\times 3$，解得 $x=5$，选 B。

16	16	【题干】：在一次作家研讨会上，男性的人数是女性的一半，总计为 24 人。 【问题】：女性有多少人？ 【解析】：根据题意：设男性的人数为 x，则女性的人数为 $2x$。可得 $x+2x=24$，解得 $x=8$，因此女性人数为 $2x=16$。
17	$\dfrac{1}{8}$	【题干】：如果 (x, y) 是以下方程组的解，且 y 不为 0。 【问题】：$(y-x)^3$ 的值为多少？ 【解析】：根据题意，将式 2 代入式 1 中：$(-3y)^2-y^2=y$，展开后得到 $y\times(8y-1)=0$，解得 $y=\dfrac{1}{8}$。再将 y 值代入式 1 中解得 $x=-\dfrac{3}{8}$，因此得到 $y-x=\dfrac{1}{2}$，$(y-x)^3$ 的值为 $\dfrac{1}{8}$。
18	5 或 7	【题干与问题】：求下列式子的一个解。 【解析】：根据题意，式子两边同时乘以 $4(x+1)(x-1)$ 得到： $$\dfrac{24(x+1)(x-1)}{x+1}-\dfrac{12(x+1)(x-1)}{x-1}=\dfrac{4(x+1)(x-1)}{4}$$ 化简后得到 $12x-36=x^2-1$，解得 $x=5$ 或者 $x=7$。
19	100	【题干】：一家银行向其商业客户提供两种不同的账户核对月费用缴纳方案。在方案 A 中，账户持有人在每笔核对服务中需要支付 \$0.15，之后就无需再支付月服务费。在方案 B 中，账户持有人需要支付月服务费 \$10.00，并且还需在每笔核对服务中支付 \$0.05 的费用。 【问题】：每月中需要有多少笔核对服务，才能使得两种方案的费用是相同的？ 【解析】：设每月的核对服务为 x 笔，则方案 A 的费用为 $0.15x$，方案 B 的费用为 $10+0.05x$。根据题意可得：$0.15x=10+0.05x$，解得 $x=100$。
20	182	【题干】：此表显示的是 P，Q，R 和 S 四个城市间的单向开车距离，例如 P 和 Q 之间的单向开车距离为 144 英里。如果 S 和 Q 之间的往返距离比 S 和 R 之间的远 16 英里，且 S 和 R 之间的往返距离比 S 和 P 之间的少 24 英里。 【问题】：X 的值为多少？ 【解析】：根据题意可得：S 与 P 之间的单向开车距离为 186 英里，则往返距离为 372 英里。S 与 R 之间的往返距离为 $372-24=348$。同理 S 与 Q 之间的往返距离为 $348+16=364$ 英里。最后注意表中给出的是单向开车距离，因此 $X=\dfrac{364}{2}=182$ 英里。
		Section 4
1	B	【题干与问题】：方程 $x^2-8x=-16$ 有几个解？ 【解析】：根据题意，将原方程化简为 $x^2-8x+16=0$，即为 $(x-4)(x-4)=0$，解得 $x=4$。因此原方程有一个解，选 B。
2	B	【题干与问题】：数列 50，44.5，39，33.5…的第 10 项是多少？ 【解析】：根据题意，观察该数列是以 5.5 递减的，因此可知公差为 -5.5，首项为 50，代入数列的公式可得第 10 项 $a_{10}=a_1+(10-1)\times-5.5=0.5$，选 B。

3	A	【题干与问题】：当 $a<-1$ 时，下面哪一个图表示的是 $y-1=\dfrac{a}{a+1}x$ 的图像？ 【解析】：根据题意，将原式改写为 $y=\dfrac{a}{a+1}x+1$，可以看出该函数的图像在 y 轴上的交点为（0，1）。再由条件 $a<-1$ 可得 $\dfrac{a}{a+1}>1$，说明该函数的斜率为正。满足上述条件的图像只有 A 项，选 A。
4	C	【题干与问题】：如果五个正整数的平均值为 70，则它们的中位数最大可能为多少？ 【解析】：根据题意可得，中位数是该五个整数从小到大排列后处于中间位置的那个数，如果要使得中位数最大，则需要满足的条件有两个： 1）该中位数之前的数要相等并且尽可能的小； 2）该中位数及其之后的数要相等并且尽可能的大。 由原题的条件可得五个数的和为 70×5＝350。由于需要该中位数之前的数相等并尽可能的小，所以取最小的正整数 1。而同时又需要该中位数及其之后的数相等并且尽可能的大，所以后面的三个数为 $\dfrac{350-2}{3}=116$，即这五个数为：1，1，116，116，116，它们的中位数为 116，选 C。
5	C	【题干】：该散点图表示的是一组 20 条鲅鱼的体长、体重以及相应的最适线。 【问题】：根据这条最适线，下面哪一个值与一条体长为 95 厘米的鲅鱼其体重最接近？ 【解析】：根据题意，从 x 轴上找到 95，再根据最适线读出相应的 y 轴上的值，可得 8.3 最接近，选 C。
6	D	【题干与问题】：下面哪一个式子最能表述每条鲅鱼的体重和体长之间的关系？ 【解析】：根据题意，要想求出这条最适线，只要找出通过该直线的两个点的坐标，就可以得到这条最适线的方程。从图中读出，这条最适线通过（60，3）和（100，9）这两点。因此该最适线的斜率为 $\dfrac{9-3}{100-60}=\dfrac{3}{20}$，所以这条最适线的方程可以写成 $w=\dfrac{3}{20}l+k$，再将点（60，3）的坐标代入得到 $k=-6$，因此方程为 $w=\dfrac{3}{20}l-6$，选 D。
7	D	【题干】：Li 夫人的第一期代数课有 25 个学生。周一有 5 位学生缺席，剩下的 20 位学生参加了一个测试。这些学生的平均成绩为 86 分。第二天，当这 5 位学生也参加了这个测试后，班级的平均成绩为 88 分。 【问题】：这 5 位学生的平均成绩是多少？ 【解析】：根据题意可得：20 位学生的总成绩为 20×86＝1,720，25 位学生的总成绩为 25×88＝2,200。这说明后来参加测试的 5 位学生的总成绩为 2,200－1,720＝480，因此他们的平均成绩为 $\dfrac{480}{5}=96$，选 D。
8	A	【题干】：对所有实数 x 而言有式子 $g(3-x)=x^2+x+1$。 【问题】：$g(7)$ 的值为多少？ 【解析】：根据题意可得：如果 $7=3-x$，则 $x=-4$。因此 $g(7)=g(3-(-4))=(-4)^2+(-4)+1=16-4+1=13$，选 A。
9	B	【题干】：对于所有 x 来说，有式子 $g(x+1)=x^2+2x+4$。 【问题】：下面哪一个式子等于 $g(x)$？ 【解析】：根据题意，将 $x-1$ 代入到 x^2+2x+4 得到 x^2+3，选 B。

题号	答案	内容
10	D	【题干】：如图 5-1 所示，矩形 ABCD 内切于圆心为 O 的圆内。 【问题】：圆的面积是多少？ 【解析】：根据题意，将 B 和 D 两点连线，得到线段 BD。根据圆内切矩形的特征可知线段 BD 过圆心，即线段 BD 为圆的直径。△BCD 是直角三角形，根据勾股定理可以得到线段 BD 的长度为 26。因此圆的半径为 13，圆的面积为 169π，选 D。 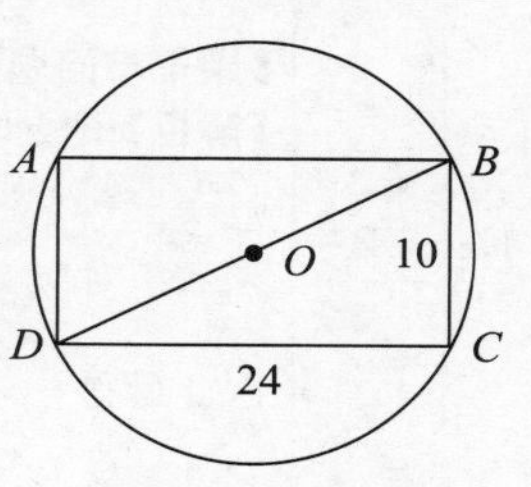 图 5-1
11	A	【题干】：一个球与一个圆锥体积相等，圆锥半径是球半径的两倍。 【问题】：该圆锥高与半径的比例是多少？ 【解析】：根据题意设球的半径为 r，圆锥的半径为 R，可得球的体积为 $V_{球} = \frac{4}{3} \times \pi \times r^3$，圆锥的体积为 $V_{圆锥} = \frac{1}{3} \times \pi R^2 \times h$，因此可得 $\frac{4}{3} \times \pi \times r^3 = \frac{1}{3} \times \pi R^2 \times h$。再将 $R = 2r$ 代入得到 $h = r = \frac{1}{2}R$。可见圆锥高与半径的比例是 0.5 : 1，选 A。
12	A	【题干】：该饼图表示的是四位销售员 Maria，Eli，Georgia 和 Zoe 在一个社交媒体网站上的月广告销售额。在这个月中，Maria 的销售额为总销售额的 25%，Eli 的销售额为 \$3,000，Georgia 的销售额为 \$5,000，Zoe 的销售额为 \$10,000。 【问题】：哪一个部分代表的是 Georgia 的销售额？ 【解析】：根据题意，由于 Maria 的销售额为总销售额的 25%，因此在图上 Maria 的销售额所对应的那部分其圆心角为 360°×25% = 90°，对应图上的 D 部分。因此，剩下三个人的销售额对应的就是图上 A、B 和 C 三部分。从题中可以看出 Georgia 的销售额高于 Eli 但低于 Zoe 的，因此取 A、B 和 C 三部分中面积中等的那块，即为 A 部分，选 A。
13	B	【题干】：详见 12 题题干。 【问题】：四位销售员的总销售额为多少？ 【解析】：根据题意，Maria 的销售额为总销售额的 25%，所以剩下三人的销售额为总销售额的 75%，对应的额度为 \$3,000 + \$5,000 + \$10,000 = \$18,000。总销售额×75% = \$18,000，得到总销售额为 \$24,000，选 B。
14	D	【题干】：如果 Eli 和 Georgia 可获得其销售额 10% 的提成，Maria 和 Zoe 可获得其销售额 15% 的提成。 【问题】：Maria 每月比 Georgia 多提成多少？ 【解析】：根据题意可得 Maria 的销售额为 \$24,000×25% = \$6,000，提成为 \$6,000×15% = \$900。Georgia 的提成为 \$5,000×10% = \$500。因此 Maria 每月比 Georgia 多提成 \$900 − \$500 = \$400，选 D。
15	C	【题干与问题】：下面哪一个函数的图像在 xy 平面坐标系中与 x 轴三次相交？ 【解析】：根据题意，函数的图像与 x 轴相交即在该交点 y 值为 0。因此原题就转化为问哪一个函数满足 $f(x) = 0$ 有三个不同的解。通过比较四个选项可以发现 C 选项满足条件，有 $x = 0$，-1 和 1 三个不同的解，选 C。
16	C	【题干】：一个网站在 7 月有 2,100 位访客访问，其中既包含注册用户也包含非注册用户。如果注册用户与非注册用户的比例为 2 : 5。 【问题】：在 7 月非注册用户比注册用户多多少人？ 【解析】：根据题意，设注册用户与非注册用户分别为 $2x$ 和 $5x$，因此一共有 $7x$ 人，即 $7x = 2,100$，解得 $x = 300$。因此注册用户为 $2x = 600$ 人，非注册用户为 $5x = 1,500$ 人，非注册用户比注册用户多 1,500 − 600 = 900 人，选 C。

17	B	【题干与问题】：当 k 为多少的时候，两个函数的图像不相交？ 【解析】：根据题意可得：如果两条线的斜率相等，说明两条线平行，那就不会相交，因此首先需要分别计算出两个直线方程的斜率。 函数 1 对应的直线方程为：$y = -\dfrac{3}{4}x - \dfrac{5}{4}$，函数 2 对应的直线方程为：$y = -\dfrac{k}{6}x - \dfrac{7}{6}$，因此可得 $-\dfrac{3}{4} = -\dfrac{k}{6}$，所以得到 $k = 4.5$，选 B。
18	C	【题干与问题】：如果 $\dfrac{x^2+1}{2} + \dfrac{x}{p} = 1$，对于所有可能的 x，下面哪一个式子是用 p 的形式来表示 x？ 【解析】：根据题意，将原式左侧的分母去掉，转换为：$p(x^2+1) + 2x = 2p$，进一步整理得 $px^2 + 2x - p = 0$。再根据一元二次方程根的求解公式 $x = \dfrac{-b \pm \sqrt{b^2 - 4ac}}{2a}$，在本题中 $a = p$，$b = 2$，所以得到 $x = \dfrac{-2 \pm \sqrt{4+4p^2}}{2p}$，选 C。
19	B	【题干】：以上两列给出的是 Pearlman 先生所教 A 和 B 两个班的测试成绩。 【问题】：下面有关于每个班级标准偏差的比较，哪一项是正确的？ 【解析】：根据题意可得：标准偏差表示的是一组数据与其平均值的偏离程度。从给出的数据可以看出 A 班的成绩从 68 到 99，而 B 班的成绩主要集中于 85～90，可见 B 班的成绩与其平均值的偏离程度更小一些，即标准偏差更小，选 B。
20	B	【题干】：有函数 $f(x) = x + 2$，满足 $f(g(1)) = 6$。 【问题】：$g(x)$ 的表达式为哪一个？ 【解析】：根据题意可得：$f(g(1)) = 6$，即将 $x = g(1)$ 代入到 $f(x) = x + 2$，得到 $g(1) + 2 = 6$，解得 $g(1) = 4$。再找当 $x = 1$ 时，四个选项中哪一个式子满足计算结果为 4，符合条件的为 $x + 3$，选 B。
21	D	【题干】：一个圆柱形的管子，厚度不计，放置在一个长方形盒子中，该盒子的尺寸分别是 3 英寸、4 英寸和 8 英寸（如图 5-2 所示）。如果该管子正好放置在从底面的左角拐角处一直到顶面的右角拐角处。 【问题】：该管子的长度与下面哪一个数字最接近？ 【解析】：根据题意，设管子放置的底面的左角拐角处为点 B，顶面的右角拐角处为点 A，点 B 在底面对角线对着的点为 C，管子的长度为 d。如图所示。 原题条件给出的盒子的长、宽和高分别为 $l = 4$，$w = 3$，$h = 8$。因此在 $\triangle ABC$ 中 $d^2 = l^2 + w^2 + h^2 = 4^2 + 3^2 + 8^2$，解得 $d = \sqrt{89} \approx 9.4$，选 D。 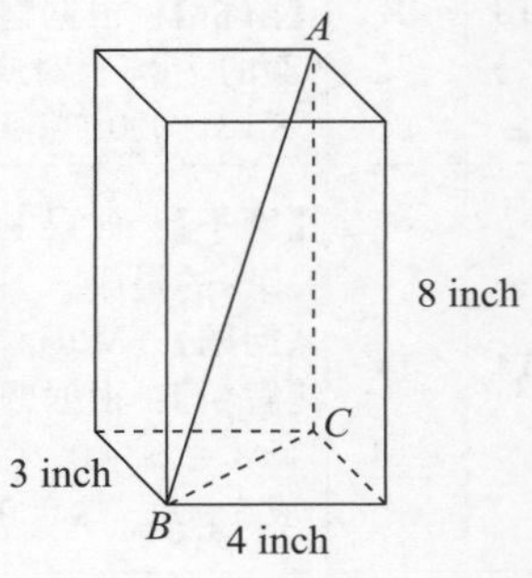
22	B	【题干】：对于所有实数，定义 f 为函数 $f(x) = 2 - |x-4|$。 【问题】：f 的最大值为多少？ 【解析】：根据题意，要想得到 f 的最大值，需要 $|x-4|$ 最小，而当 $x = 4$ 的时候 $|x-4|$ 的值最小为 0，所以 f 的最大值为 $2 - 0 = 2$，选 B。

23	B	**【题干】**：地平线度假酒店推出了单人房，每晚 $150，一次代客泊车费用为 $35。该州征房价税为 6.5%，代客泊车不征税。 **【问题】**：下面哪一个式子表示地平线度假酒店住宿单人房 n 个晚上所收取的总费用 c？其中包括代客泊车费用和税费。 **【解析】**：根据题意可得：n 个晚上的房费是 $150 \times n$，税后为 $(1 + 6.5\%) \times 150 \times n$，再加上代客泊车费用 $35，共计 $1.065 \times 150n + 35$，选 B。
24	D	**【题干与问题】**：如果 y 的值与 x 的值成正比，当 $x = c$ 的时候 $y = 12$，则当 $x = 8$ 的时候，用 c 来表示 y 的值为多少？ **【解析】**：根据题意 y 的值与 x 的值成正比，因此可得 $\frac{12}{c} = \frac{y}{8}$，化简后可得 $y = \frac{96}{c}$，选 D。
25	B	**【题干】**：不同类别人群的票价如表所示。有四种票的类型：成人票、儿童票、老年票和学生票，同时表中也给出了对应票种的总收入。 **【问题】**：根据表所提供的信息，一张老年票多少钱？ **【解析】**：根据题意可得：老年票共计 16 张，总收入为 $96，因此每张价格为 $6，选 B。
26	D	**【题干】**：详见 25 题题干。 **【问题】**：一张成人票比一张学生票贵多少钱？ **【解析】**：根据题意可得：成人票共计 84 张，总收入为 $630，因此每张价格为 $7.5。学生票共计 110 张，总收入为 $495，因此每张价格为 $4.5。可见一张成人票比一张学生票贵 $7.5 − $4.5 = $3.0，选 D。
27	C	**【题干】**：详见 25 题题干。 **【问题】**：下面哪一个值最接近平均票价？ **【解析】**：根据题意可得：一共是 250 张票，总收入为 $1,421，因此平均的票价为 $\frac{1,421}{250} = 5.684 \approx 5.68$，选 C。
28	B	**【题干与问题】**：下面哪一项不等于 $i^6 - i^2$？ **【解析】**：根据题意可得：$i^6 - i^2 = (i^2)^3 - (-1) = -1 - (-1) = 0$。A 项：$i^5 - i = i - i = 0$；B 项：$i^4 = 1$；C 项：$2i^3 + 2i = -2i + 2i = 0$；D 项：$1 + i^6 = 1 + (-1) = 0$，选 B。
29	D	**【题干与问题】**：如果 $\sin y = \frac{a}{b}$，且 $0 < y < \frac{\pi}{2}$，下面哪一个式子表示的是 $\sin\left(\frac{\pi}{2} - y\right)$？ **【解析】**：根据题意可得一个直角三角形 ABC，如图 5-3 所示。由 $\sin y = \frac{BC}{AC} = \frac{a}{b}$ 可得 $AC = b$，$BC = a$，$AB = \sqrt{b^2 - a^2}$。根据三角函数的性质 $\sin\left(\frac{\pi}{2} - y\right) = \cos y = \frac{AB}{AC} = \frac{\sqrt{b^2 - a^2}}{b}$，选 D。 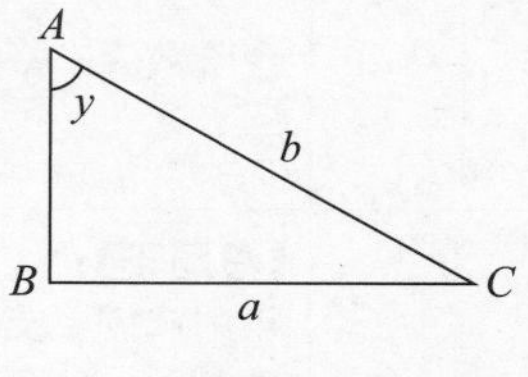 图 5-3
30	B	**【题干与问题】**：如果 $f(x) = (x^2)^{-2b}$ 且 $f(3) = 3$，问 b 的值是多少？ **【解析】**：根据题意可得，$f(3) = (3^2)^{-2b} = 3^{-4b} = 3$，所以 $-4b = 1$，$b = -\frac{1}{4}$，选 B。

31	720	【题干与问题】：Fred 收藏的贝壳数量是 Phil 的 80%，如果 Phil 的贝壳比 Fred 多 80 个，那么他们两个一共有多少个贝壳？ 【解析】：设 Phil 的贝壳数为 P，则 Fred 的贝壳数为 $0.8 \times P$。根据题意可得：$P - 0.8P = 0.2P = 80$，所以 $P = 400$。因此 Phil 的贝壳数为 400，Fred 的贝壳数为 320，两人一共有 $400 + 320 = 720$ 个。
32	3	【题干与问题】：如果有函数 $f(x) = 3 + \dfrac{5}{x}$，则 $f(x)$ 的值不可能为多少？ 【解析】：根据题意可得：对于 x 而言，$\dfrac{5}{x} \neq 0$，因此 $f(x) \neq 3$。
33	7	【题干与问题】：如果有等式 $\dfrac{x}{x+1} + \dfrac{1}{x-1} = \dfrac{25}{24}$，且 $x > 0$，则 x 的值为多少？ 【解析】：根据题意，等式两边都乘上 $24(x+1)(x-1)$，可得 $24x(x-1) + 24(x+1) = 25(x+1)(x-1)$，整理得 $x^2 - 49 = 0$，因此 $x = \pm 7$，同时需要满足原题条件 $x > 0$，因此得到 $x = 7$。
34	2	【题干与问题】：如果直线 $y = -4x - 3$ 与直线 $y = -3x - b$ 相交于点 $(-1, c)$，则 b 的值为多少？ 【解析】：根据题意，可将点的坐标代入到两个直线方程中去： 1) 代入 $y = -4x - 3$ 可得 $c = -4 \times (-1) - 3 = 1$； 2) 代入 $y = -3x - b$ 可得 $c = 1 = 3 - b$，因此 $b = 2$。
35	15	【题干】：一位农民从相邻的农场购买了若干只动物：其中 6 只动物每只 \$100，10 只动物每只 \$200，k 只动物每只 \$400。如果 k 是一个正的奇整数，而且所有动物价格的中位数为 \$200。 【问题】：k 的值最大可能为多少？ 【解析】：根据题意，我们可以将知道的数据进行从小到大排列： 100, 100, 100, 100, 100, 100, 200, 200, 200, 200, 200, 200, 200, 200, 200, 200, 400, ⋯ 由于不知道 k 是多少，要想满足动物价格的中位数为 200，且 k 值要尽可能地最大，因此只要满足上面数列中最后一个"200"是处在中间位置即可。从上面的数列中可知最后一个"200"之前共有 15 个数字，因此 $k = 15$。
36	5.25	【题干】：如图所示，一个楼梯每一级台阶的宽为 0.25 米，高为 0.2 米。图中所有的角度都是直角，楼梯的高度为 3.6 米，楼梯顶端的平台宽 1 米。 【问题】：图中 AB 的长度是多少？ 【解析】：根据题意可得，每级台阶的高度是 0.2 米，楼梯的高度为 3.6 米，可得楼梯的总台阶数为 $\dfrac{3.6}{0.2} = 18$。再根据楼梯顶端的平台长度为 1 米，所以 $AB = 1 + 0.25 \times (18 - 1) = 5.25$。
37	47	【总题干】：一家网络服务供应商为居民用户提供三种不同类型的套餐。在 A 套餐中，第一年用户需要支付的费用为 \$500，之后每月支付 \$80。在 B 套餐中，用户每月支付的费用为 \$68。C 套餐是"高网速"套餐，每月支付 \$92，网速则可以提高 200%。 【题干】：Isabelle 使用 A 套超过 1 年了，她现在反思一下她所使用的套餐情况，发现如果同样的时间里使用了 B 套餐，她的网费一共可以省下 \$104。 【问题】：到她开始反思的时候，Isabelle 使用 A 套餐有多长时间了？ 【解析】：根据题意，设 Isabelle 使用了 A 套餐的时间为 y 个月，因此可以得到总费用为 $500 + 80 \times (y - 12)$。如果同样的时间段使用 B 套餐，则费用应该为 $68 \times y$。根据原题所给的条件：$500 + 80 \times (y - 12) - 68 \times y = 104$，解得 $y = 47$。

38	33	**【题干】**：由于家庭商务需要，Isabelle 现在考虑更换 B 套餐或者是 C 套餐。但是她经过计算发现高网速套餐每月最多可以帮她节省 45 分钟的工作时间。 **【问题】**：当每小时工资率（工资成本）最少为多少的时候，以美元/小时来表示，Isabelle 选择 C 套餐更有意义？（取整数） **【解析】**：根据题意可得，每个月 C 套餐的成本比 B 套餐成本高 92－68＝24，每个月因为使用 C 套餐而节省的时间为 45 分钟。因此设每小时工资率为 r，Isabelle 每个月如果选 C 套餐，则多支付 \$24，但可以少付 45 分钟所对应的工资。因此存在一个临界条件，就是选 C 套餐多出的成本＝选 C 套餐后节省的工资数，即：$24 = r \times \dfrac{45}{60}$，解得 $r = 32$。因此当 $r > 32$ 的时候，选 C 套餐后节省的工资数大于选 C 套餐多出的成本，是有意义的。最终根据题意选择符合条件的最小数值，为 33。

Test 2

Section 3

Math Test—No Calculator
25 MINUTES, 20 QUESTIONS

Turn to Section 3 of your answer sheet to answer the questions in this section.

DIRECTIONS

For questions 1—15, solve each problem, choose the best answer from the choices provided, and fill in the corresponding circle on your answer sheet. For questions 16—20, solve the problem and enter your answer in the grid on the answer sheet. Please refer to the directions before question 16 on how to enter your answers in the grid. You may use any available space in your test booklet for scratch work.

NOTES

1. The use of a calculator is not permitted.
2. All variables and expressions used represent real numbers unless otherwise indicated.
3. Figures provided in this test are drawn to scale unless otherwise indicated.
4. All figures lie in a plane unless otherwise indicated.
5. Unless otherwise indicated, the domain of a given function f is the set of all real numbers x for which $f(x)$ is a real number.

REFERENCE

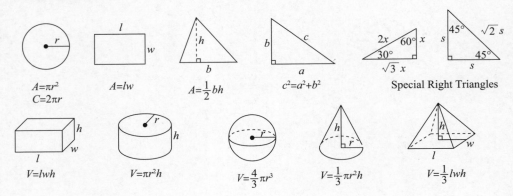

The number of degrees of arc in a circle is 360.
The number of radians of arc in a circle is 2π.
The sum of the measures in degrees of the angles of a triangle is 180.

In the *xy*-coordinate plane, what is the area of the rectangle with opposite vertices at $(-3, -1)$ and $(3, 1)$ in units squared?
A. 3 B. 6 C. 9 D. 12

In a set of five positive whole numbers, the mode is 90 and the average (arithmetic mean) is 80. Which of the following statements is false?
A. The number 90 appears two, three, or four times in the set
B. The number 240 cannot appear in the set
C. The number 80 must appear exactly once in the set
D. The five numbers must have a sum of 400

A square with an area of 25 is changed into a rectangle with an area of 24 by increasing the width and reducing the length. If the length was reduced by 2, by how much was the width increased?
A. 2 B. 3 C. 4 D. 5

The following Venn diagram (Fig. 5-9) shows the ice-cream flavor choice of 36 children at an ice-cream party. Each child could choose vanilla ice cream, chocolate ice cream, both, or neither. What percent of the children had chocolate ice cream only?
A. 10%   B. 25%
C. 50%   D. 75%

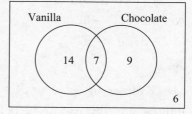

Fig. 5-9

When an object is dropped, the number feet N that it falls is given by the formula $N = \frac{1}{2}gt^2$, where t is the time in seconds since it was dropped and g is 32.2. If takes 5 seconds for the object to reach the ground, how many feet does it fall during the last 2 seconds?
A. 64.4   B. 96.6   C. 161.0   D. 257.6

The distance a free falling object has traveled can be modeled by the equation $d = \frac{1}{2}at^2$, where a is acceleration due to gravity and t is the amount of time the object has fallen. What is t in terms of a and d?

A. $t = \sqrt{\dfrac{da}{2}}$ B. $t = \sqrt{\dfrac{2d}{a}}$ C. $t = \left(\dfrac{da}{2}\right)^2$ D. $t = \left(\dfrac{2d}{a}\right)^2$

In the figure (Fig. 5-10), points P, Q, R, S, and T lie on the same line, and R is the center of the large circle. If the three smaller circles are congruent and the radius of the large circle is 6, what is the radius of one of the smaller circles?

A. 1 B. 2
C. 3 D. 4

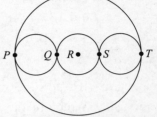

Note: Figure not drawn to scale
Fig. 5-10

Which of the following points is NOT on the graph of the line $-2x - 3y = 36$ in the xy-plane?

A. $(-9, 6)$ B. $(-24, 4)$ C. $(6, -16)$ D. $(12, -20)$

During a coyote repopulation study, researchers determine that the equation $P = 250(1.32)^t$ describes the population P of coyotes t years after their introduction into a new region. Which of the following gives the values of I, the initial population of coyotes, and r, the annual percent increase in this population?

A. $I = 250$, $r = 32\%$ B. $I = 250$, $r = 132\%$
C. $I = 330$, $r = 32\%$ D. $I = 330$, $r = 132\%$

Mrs. Krauser invested a part of her $6,000 inheritance at 9 percent simple annual interest and the rest at 12 percent simple annual interest. If the total interest earned in one year was $660, how much did she invest at 12 percent?

A. 2,000 B. 3,000 C. 4,000 D. 5,000

Barbara and Marc each rolled a single die 50 times. The frequency distributions for each of them are given in the tables (Table 5-4 and Table 5-5).

Table 5-4   Distribution of Barbara's 50 Rolls

Number	1	2	3	4	5	6
Frequency	4	10	8	8	8	12

Table 5-5 Distribution of Marc's 50 Rolls

Number	1	2	3	4	5	6
Frequency	7	11	13	3	8	8

If the two distributions are combined into a single frequency distribution representing all 100 rolls, what is the median value of those 100 rolls?
A. 3   B. 3.25   C. 3.5   D. 4

$$[(2x+y)+(x+2y)]^2$$

Which of the following expressions is equivalent to the expression above?
A. $3(x^2+y^2)$ B. $9(x^2+y^2)$ C. $3(x+y)^2$ D. $9(x+y)^2$

To get to a business meeting, Joanna drove m miles in h hours, and arrived $\frac{1}{2}$ hour early. At what rate should she have driven to arrive exactly on time?

A. $\frac{2m+h}{2h}$ B. $\frac{2m-h}{2h}$ C. $\frac{2m}{2h-1}$ D. $\frac{2m}{2h+1}$

Two cylindrical tanks have the same height, but the radius of the larger tank equals the diameter of the smaller tank. If the volume of the larger tank is $k\%$ more than the volume of the smaller tank, what is the value of k?
A. 100   B. 200   C. 300   D. 400

15

How many points of intersection are there of the graphs whose equations are $y=-(x-3)^2+3$ and $y=(x+3)^2-3$?
A. 0   B. 1   C. 2   D. More than 2

DIRECTIONS

For questions 16—20, solve the problem and enter your answer in the grid, as described below, on the answer sheet.
1. Although not required, it is suggested that you write your answer in the boxes at the top of the columns to help you fill in the circles accurately. You will receive credit only if the circles are filled in correctly.
2. Mark no more than one circle in any column.
3. No question has a negative answer.
4. Some problems may have more than one correct answer. In such cases, grid only one answer.

SAT-1 数学轻松突破 800 分：思路与技巧的飞跃

5. Mixed numbers such as $3\frac{1}{2}$ must be gridded as 3.5 or 7/2. (If is entered into the grid, it will be interpreted as $\frac{31}{2}$, not as $3\frac{1}{2}$.)

6. Decimal Answers: If you obtain a decimal answer with more digits than the grid can accomodate, it may be either rounded or truncated, but it must fill the entire grid.

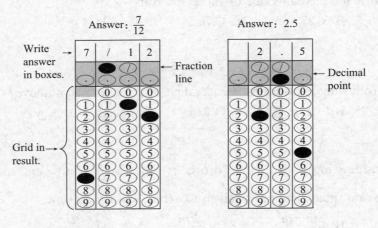

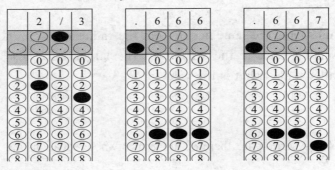

Answer: 201—either position is correct

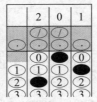 

NOTE: You may start your answers in any column, space permitting. Columns you don't need to use should be left blank.

16

If $i = \sqrt{-1}$ and $(1+i) \div (1-i) = a + bi$, where a and b are real numbers, what are the values of $a + b$?

The table (Table 5-6) shows a set of ordered pairs that correspond to the function $h(x) = \dfrac{x^2}{2} + k$. What is the value of k?

Table 5-6

x	$h(x)$
3	6
5	14

In the figure (Fig. 5-11), if $AB = BD = DF = FH = 1$, what is the ratio of the area of trapezoid *FHJG* to the area of trapezoid *BCED*?

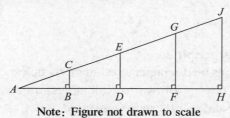

Note: Figure not drawn to scale

Fig. 5-11

In the figure (Fig. 5-12), *ABCD* is a square with side of length 2. If *E* is the midpoint of line segment *AB* and *F* is the midpoint of line segment *AD*, what is the area of quadrilateral *CFAE*, in units squared?

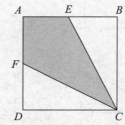

Note: Figure not drawn to scale

Fig. 5-12

If the lines of $x + 2y = 7$ and $2x - ky = 5$ are perpendicular, what is the value of k?

STOP

If you complete this section before the end of your allotted time, check your work on this section only. Do NOT use the time to work on another section.

Section 4

Math Test—Calculator

55 MINUTES, 38 QUESTIONS

Turn to Section 4 of your answer sheet to answer the questions in this section.

DIRECTIONS

For questions 1—30, solve each problem, choose the best answer from the choices provided, and fill in the corresponding circle on your answer sheet. For questions 31—38, solve the problem and enter your answer in the grid on the answer sheet. Please refer to the directions before question 31 on how to enter your answers in the grid. You may use any available space in your test booklet for scratch work.

NOTES

1. The use of a calculator is permitted.
2. All variables and expressions used represent real numbers unless otherwise indicated.
3. Figures provided in this test are drawn to scale unless otherwise indicated.
4. All figures lie in a plane unless otherwise indicated.
5. Unless otherwise indicated, the domain of a given function f is the set of all real numbers x for which $f(x)$ is a real number.

REFERENCE

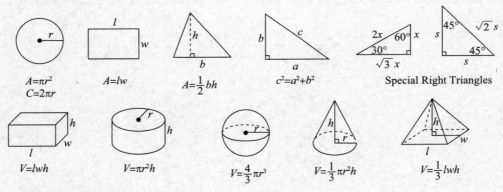

The number of degrees of arc in a circle is 360.

The number of radians of arc in a circle is 2π.

The sum of the measures in degrees of the angles of a triangle is 180.

CONTINUE

If $|10-3y|<3$, which of the following is a possible value of y?
A. 0
B. 1
C. 2
D. 3

Three cars drove past a speed-limit sign on a highway. Car A was traveling twice as fast as Car B, and Car C was traveling 20 miles per hour faster than Car B. If Car C was traveling at 60 miles per hour, how fast was Car A going?
A. 20 miles per hour
B. 30 miles per hour
C. 40 miles per hour
D. 80 miles per hour

If $-1<x<0$, which of the following statements must be true?

Ⅰ. $x>\dfrac{x}{2}$

Ⅱ. $x^2>x$

Ⅲ. $x^3>x^2$

A. Ⅰ only
B. Ⅱ only
C. Ⅰ and Ⅱ only
D. Ⅰ, Ⅱ, and Ⅲ

The estate of a wealthy man was distributed as follows: 10% to his wife, 5% divided equally among his three children, 5% divided equally among his five grandchildren, and the balance to a charitable trust. If the trust received $1,000,000, how much did each grandchild inherit?
A. $10,000
B. $12,500
C. $20,000
D. $62,500

The bar graph (Fig. 5 – 13) shows the number of students in four universities who received financial aid from the university in 2015. The average size of the financial aid package per student at universities A, B, C, and D was $15,500, $21,000, $18,700, and $14,300, respectively. Which university gave out the greatest total amount of financial aid?
A. A
B. B
C. C
D. D

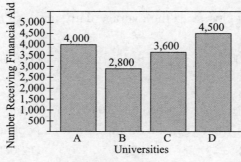

Fig. 5 – 13

6

Marie has a website where she sells CDs and DVDs. She purchases her CDs for $2.75 each and her DVDs for $5.75 each. Marie pays $0.95 to mail each CD and DVD to her customers. She charges $4.99 per CD and $9.99 per DVD plus a postage and handling fee of $1.75 per CD or DVD. Which of the following represents her profit, P, in dollars, on the sale of x CDs and y DVDs?

A. $P = 3.04x + 5.04y$
B. $P = 2.24x + 4.24y + 0.80$
C. $P = 2.24x + 4.24y + 0.80xy$
D. $P = 3.00(x + y) + 0.80(x + y)$

7

A survey of 500 registered voters in a certain state was taken to ascertain the number of Democrats, Republicans, and Independents who supported a certain ballot initiative called Proposition 8. The results of that survey are tabulated below (Table 5-7).

Table 5-7

	Support Proposition 8	Opposed to Proposition 8	Undecided	Total
Democrats	113	32	40	185
Republicans	35	145	30	210
Independents	44	41	20	105
Total	192	218	90	500

On Election Day, all of the voters in the survey who had expressed support for the proposition voted for it and all of the voters who had been opposed to the proposition voted against it. If in addition, 80% of those who had been undecided voted for the proposition and 20% voted against it, what percent of the 500 people in the survey voted for the proposition?

A. 42.7%   B. 47.2%   C. 52.8%   D. 58.2%

Questions 8 and 9 refer to the following information.

According to the United States Census Bureau, on average there is a birth in the United States every 8 seconds, a death every 12 seconds, and a net increase of one person due to immigration and emigration every 30 seconds.

8

Which of the following is closest to the average daily increase in the population of the United States?

A. 2,280   B. 4,260   C. 6,480   D. 9,520

9

The population of the United States reached 320,000,000 in January of 2015. According to the Census Bureau's analysis, in what year should the country's population reach 350,000,000?

A. 2017   B. 2022   C. 2028   D. 2032

The diagram (Fig. 5-14) represents a conical tank whose radius is 3 feet and whose height is 6 feet. If the tank is full of water and if exactly half the water in the tank is poured out, what is the height, to the nearest inch, of the water remaining in the tank?

A. 36 B. 48
C. 54 D. 57

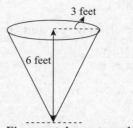

Note: Figure not drawn to scale
Fig. 5-14

In 2000, Jennifer invested $1,000 in a seven-year Certificate of Deposit (CD) that paid 2% interest, compounded annually. When that CD matured in 2007, she invested all of the money in another seven-year CD, also paying 2% compounded annually, that matured in 2014. To the nearest dollar, how much more money did Jennifer earn from 2007 to 2014 than she did from 2000 to 2007?

A. $22 B. $44 C. $149 D. $171

The function f is defined by $f(x) = x^2 + ax + a$, where a is a constant. What is $f(5)$ in terms of a?

A. $25 + a$ B. $25 + 2a$ C. $5 + a^2$ D. $25 + 6a$

What is the solution set of this system of equations?

$$y - x = 3$$
$$x^2 - 7y + 31 = 0$$

A. $\{(2, 5), (5, 2)\}$ B. $\{(2, 5), (5, 8)\}$
C. $\{(5, 8), (8, 5)\}$ D. $\{(8, 5), (8, 8)\}$

Find the value of x if $\dfrac{x}{12} - \dfrac{x+2}{4} < 0$.

A. $x < -3$ B. $x > -3$ C. $x < 3$ D. $x > 3$

If $\cos\dfrac{\pi}{3} = x - 1$, then what is the value of x?

A. $\dfrac{1}{2}$ B. $\dfrac{3}{2}$

C. $\dfrac{\pi}{3} + 1$ D. x has two values

If $x \neq 0$, $y \neq 0$, and $x \neq y$, then which of the following expressions is equivalent to $\dfrac{x^{-1} - y^{-1}}{x - y}$?

A. $-\dfrac{1}{xy}$ B. $\dfrac{1}{(x-y)^2}$

C. $\dfrac{y^2 - x^2}{xy}$ D. $\dfrac{x^2 - y^2}{xy}$

A band is recording an album. They rent a studio at $200 per day, which gives them access for 12 hours. The day rate cannot be prorated, so studio time must be purchased by the day. Their sound engineer costs $28 per hour, and is needed for half the total recording time. If it will take 50 hours to complete, what is the average cost per hour over the entire course of recording?

A. $31.60 B. $34
C. $44.50 D. $48

In the semicircle (Fig. 5 - 15), the center is at (0, 0). Which of the following are the y-coordinates of two points on this semicircle whose x-coordinates are equal?

A. $y = 0, 5$ B. $y = 4, -4$
C. $y = 1, -3$ D. $y = 2, -4$

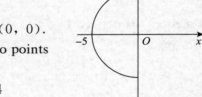

Note: Figure not drawn to scale

Fig. 5 - 15

In a sequence of evenly spaced numbers, the first term is 7, and the 20th term is 159. Which following number would be the fourth term of the sequence?

A. 32 B. 31 C. 30 D. 29

In a class of 100 students, 65 take Spanish, 32 take art, and 14 take both Spanish and art. How many students do not take either Spanish or art?

A. 3 B. 11 C. 17 D. 35

If the expression $\dfrac{5 - 9x^2}{2 - 3x}$ is written in the equivalent form $\dfrac{1}{2 - 3x} + A$, what is A in terms of x?

A. $2 - 3x$ B. $2 + 3x$ C. $9x^2$ D. 5

In the following sequence, the first term is 2, and each term after the first term is 3 less than 3 times the previous term. What is the value of k ?

$$2, 3, 6, k, 42$$

A. 10 B. 12 C. 14 D. 15

The figures (Fig. 5 – 16) show the graphs of the functions f and g. The function f is defined by $f(x) = 2|x+2|$ and the function g is defined by $g(x) = f(x+h) + k$, where h and k are constants. What is the value of $|h-k|$?

A. 1 B. 4 C. 5 D. 8

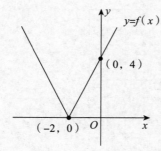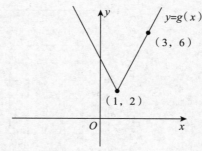

Fig. 5 – 16

Suppose that the average (arithmetic mean) of a, b, and c is h, the average of b, c, and d is j, and the average of d and e is k. What is the average of a and e?

A. $h - j + k$

B. $\dfrac{3h + 3j - 2k}{2}$

C. $\dfrac{3h - 3j + 2k}{2}$

D. $\dfrac{3h - 3j + 2k}{5}$

John pays his workers with tickets. Each ticket is worth \$32. A worker earns \$60 per hour. John has 24 tickets. Which of the following functions models the number of tickets remaining t hours after a worker has started his work?

A. $f(t) = 24 - \dfrac{60t}{32}$

B. $f(t) = 24 - \dfrac{32t}{60}$

C. $f(t) = \dfrac{24 - 60t}{32}$

D. $f(t) = \dfrac{24 - 32t}{60}$

A researcher is interested in estimating the average time a mother with one child up to 4 years old spends watching TV between 5 pm to 10 pm. She surveyed 80 mothers of the above population and found the sample mean to be 34 minutes while the margin of error for this estimate was 3.1 minutes. If she were to replicate the sampling in an attempt to get a smaller margin of error, which of the following samples will most likely result in a reduced margin of error for the mean time spent watching TV between 5 pm and 10 pm by a mother with one child up to 4 years old?

A. 50 randomly selected mothers with one child up to four years old
B. 170 randomly selected mothers with multiple children up to four years old
C. 250 randomly selected mothers with multiple children up to four years old
D. 320 randomly selected mothers with one child up to four years old

Questions 27 and 28 refer to the following graph (Fig. 5 – 17).

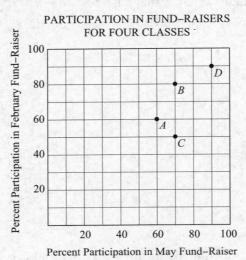

Fig. 5 – 17

Four different classes at Corbett Elementary School participated in two fund-raisers last year, one in February and another in May. The rates of participation for each class are recorded in the graph above. Which class had the greatest change in percent participation from the February fundraiser to the May fundraiser?

A. Class A B. Class B C. Class C D. Class D

If there were 20 students each in Class A and Class C, and 30 students each in Class B and Class D, how many students participated in the May fund-raiser?

A. 71 B. 72 C. 74 D. 76

$$3x^2 = 4x + c$$

In the equation above, c is a constant. If $x = -1$ is a solution of this equation, what other value of x satisfies the equation?

A. $\dfrac{1}{7}$ B. $\dfrac{4}{3}$ C. $\dfrac{7}{3}$ D. 7

In your search for a summer job, you are given the following offers.

Offer 1: At Timmy's Tacos, you will earn $4.50 an hour. However, you will be required to purchase a uniform for $45.00. You will be expected to work 20 hours each week.

Offer 2: At Kelly's Car Wash, you will earn $3.50 an hour. No special attire is required. You must agree to work 20 hours each week.

Before deciding which job offer you wish to take, you consider the factors. Which conclusion below is NOT true?

A. If I work 8 weeks at Kelly's Car Wash and save all my earnings, I'll be able to save $560

B. If I take the job at Timmy's Tacos, I'll have to work 10 hours just to pay for purchasing my uniform

C. If I only plan to work for two weeks, I should choose the job at Kelly's Car Wash

D. The job at Timmy's Tacos pays more if I work more than forty hours

---

DIRECTIONS

For questions 31—38, solve the problem and enter your answer in the grid, as described below, on the answer sheet.

1. Although not required, it is suggested that you write your answer in the boxes at the top of the columns to help you fill in the circles accurately. You will receive credit only if the circles are filled in correctly.

2. Mark no more than one circle in any column.

3. No question has a negative answer.

4. Some problems may have more than one correct answer. In such cases, grid only one answer.

5. Mixed numbers such as $3\frac{1}{2}$ must be gridded as 3.5 or 7/2. (If $\boxed{3\ 1\ /\ 2}$ is entered into the grid, it will be interpreted as $\frac{31}{2}$, not as $3\frac{1}{2}$.)

6. Decimal answers: If you obtain a decimal answer with more digits than the grid can accomodate, it may be either rounded or truncated, but it must fill the entire grid.

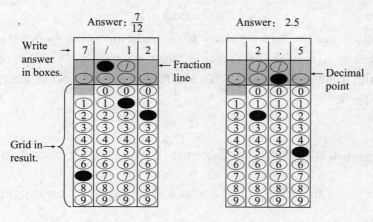

Acceptable ways to grid $\frac{2}{3}$ are:

Answer: 201–either position is correct

NOTE: You may start your answers in any column, space permitting. Columns you don't need to use should be left blank.

CONTINUE

Lauren has $80 in her savings account. When she receives her paycheck, she makes a deposit which brings the balance up to $120. By what percent does the total amount in her account increase as a result of this deposit?

If $\cos(x - \pi) = 0.4$, what is the value of $\sin^2 x$?

$$V(t) = 1,000(1 + k)^m$$

An analyst wants to use the formula above to estimate the value, in dollars, of a $1,000 initial investment in a mutual fund after m quarters have passed. If a $1,000 initial investment in this fund is worth $1,102.50 after 2 quarters, what number should the analyst choose for k?

Line l passes through the origin and is parallel to the line $y = \frac{2}{3}x - 6$. If line l intersects the line $y = \frac{1}{2}x - 4$ at the point (x, y), what is the value of the product xy?

$$hx + 4y = -3$$

The equation above is the equation of a line in the xy-plane, and h is a constant. If the slope of this line is −13, what is the value of h?

The hexagonal face of the block shown in the figure (Fig. 5−18) has sides of equal length and angles of equal measure. If each lateral face is rectangular, what is the area, in square inches, of one lateral face?

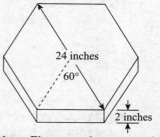

Note: Figure not drawn to scale
Fig. 5−18

Question 37 and 38 refer to the following information.
A shoe company produces batches of leather shoes and boots. Each batch is made from a starting marerial of 2,000 square feet of leather. A pair of boots requires 10 square feet of leather to produce, and a pair of shoes requires 3 square feet.

The company needs to produce a batch of exactly 500 pairs. How many pairs of boots are in this batch?

A consultant is brought in to assess the production process, and he suggest an adjustment based on market analysis. The demand for boots is on the rise, and a single pair can now be sold for $300, whereas a pair of shoes sells for $75. Assuming all other production costs are the same, how many pairs of boots should be produced in order to maximize profits?

STOP
If you complete this section before the end of your allotted time, check your work on this section only. Do NOT use the time to work on another section.

■ Test 2 答案与解析

		Section 3
1	D	【题干与问题】：在 xy 坐标平面中，以（-3，-1）和（3，1）为相对顶点的矩形面积是多少？ 【解析】：根据题意，如图 5-4 所示，该矩形的长为 6，宽为 2，因此面积为 $6×2=12$，选 D。 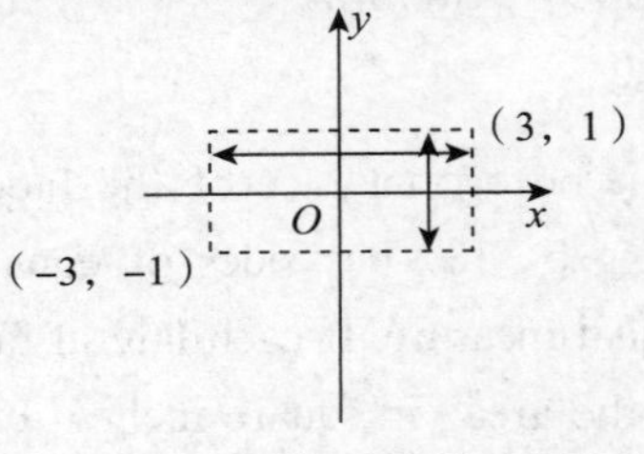 图 5-4
2	C	【题干】：一个由 5 个正整数所组成的数列，其众数为 90，平均数为 80。 【问题】：下面的表述哪一个是错误的？ 【解析】：A 项中指出 90 在该数列中出现了 2、3 或者 4 次。原题给出的条件：该数列的众数为 90，根据众数的定义，在该数列中"90"最少出现 2 次，最多出现 4 次，故 A 项正确。B 项中指出 240 不可能出现在该数列中。从原题给出的条件"平均数是 80"可得该数列 5 个数的和为 $80×5=400$（因此 D 项正确），而"90"至少出现 2 次，因此这两个数的和为 $90×2=180$，而剩余三个数的和最多为 $400-180=220$，而"240"已经大于这个和，所以"240"不可能在该数列中，B 项正确。C 项指出"80"一定在该数列中，这不一定，只要满足五个数的平均值是 80 即可，并不需要一定有"80"这个数字存在，C 项不正确，选 C。
3	B	【题干】：一个正方形的面积为 25，现在通过增加宽度和减少长度使其变换成一个面积为 24 的矩形，如果长度减少了 2。 【问题】：宽度增加了多少？ 【解析】：根据题意可得，原正方形的面积为 25，因此边长为 5，变换成矩形后，长度变为 $5-2=3$。长方形的面积为 24，因此得到宽为 8。可见宽增加了 $8-5=3$，选 B。
4	B	【题干】：上面的维恩图表示在一次冰淇淋聚会上询问了 36 个孩子对冰淇淋口味的喜好情况：喜欢香草味、巧克力味，两个口味都喜欢还是都不喜欢。 【问题】：只喜欢巧克力口味冰淇淋的孩子所占比例是多少？ 【解析】：从图中可以看出，一共有 $14+7+9+6=36$ 个孩子，其中只喜欢巧克力口味冰淇淋的孩子有 9 个，比例为 $\frac{9}{36}×100\%=25\%$，选 B。
5	D	【题干】：当一个物体下落时，下落的英尺数 N 可以用公式 $N=\frac{1}{2}gt^2$ 来表示，其中 t 是该物体的下落时间（单位：秒），g 为 32.2。 【问题】：若该物体下落 5 秒钟后到达地面，问它在最后 2 秒下落的距离是多少英尺？ 【解析】：根据题意可得前 3 秒钟该物体下落的距离为： $$N=\frac{1}{2}gt^2=\frac{1}{2}×32.2×9=144.9$$ 同理前 5 秒钟该物体下落的距离为： $$N=\frac{1}{2}gt^2=\frac{1}{2}×32.2×25=402.5$$ 因此最后的 2 秒下落的距离为前 5 秒钟该物体下落的距离减去前 3 秒钟该物体下落的距离：$402.5-144.9=257.6$，选 D。

6	B	**【题干】**：一个自由落体，其下降的距离可以用方程 $d=\frac{1}{2}at^2$ 来表示，其中 a 代表重力加速度，t 代表下落的时间。 **【问题】**：t 用 a 和 d 来表示是多少？ **【解析】**：根据题意可得 $d=\frac{1}{2}at^2$，转换得 $t^2=\frac{2d}{a}$，最后得到 $t=\sqrt{\frac{2d}{a}}$，选 B。
7	B	**【题干】**：如图 5-5 所示，点 P，Q，R，S 和 T 在一条直线上，点 R 是大圆的圆心。如果三个小圆全等，且大圆的半径为 6。 **【问题】**：小圆的半径为多少？ **【解析】**：根据题意，如图所示，将点 R 和 T 连接，得到线段 RT，RT 为大圆的半径，长度为 6，同时也等于中间小圆的半径+右侧小圆的直径。设三个全等小圆的半径为 r，因此可得 $r+2r=6$，解得小圆的半径 $r=2$，选 B。 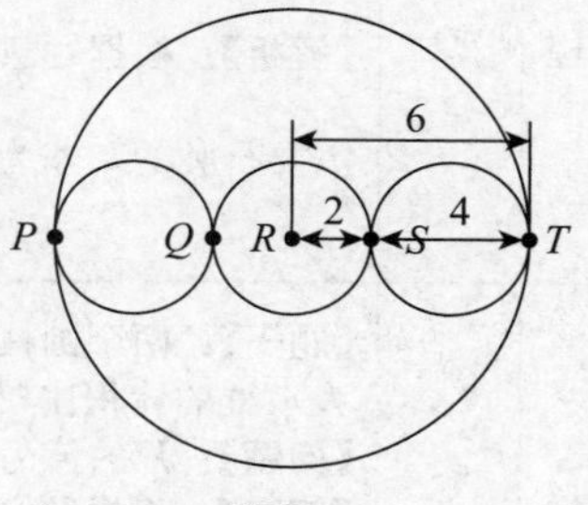 图 5-5
8	A	**【题干与问题】**：在 xy 平面坐标系中，下面哪一个点不在直线 $-2x-3y=36$ 的图像上？ **【解析】**：根据题意，可将四个点的坐标值代入直线方程中，不符合方程的即说明该点不在该直线上。 A 项：$-2\times(-9)-3\times 6=18-18=0\neq 36$； B 项：$-2\times(-24)-3\times 4=48-12=36$； C 项：$-2\times 6-3\times(-16)=-12+48=36$； D 项：$-2\times 12-3\times(-20)=-24+60=36$。 因此选 A。
9	A	**【题干】**：在一项研究北美小狼重新形成种群的研究中，研究者得到了一个方程式 $P=250(1.32)^t$，该方程用来描述北美小狼在引入到一个新的区域 t 年之后的种群数量 P。 **【问题】**：下面给出的四组数，哪一组表示的是北美小狼初始种群密度值 I 和年增长率 r？ **【解析】**：根据题意可得，初始种群密度就是指在刚引入到这个新区域时的种群密度，即 $t=0$ 时的 P 值，解得 $P=250\times(1.32)^0=250$。而过了 1 年之后，该种群的数量为 $P=250\times(1.32)^1=330$，因此年增长率为 $\frac{330-250}{250}=0.32=32\%$，选 A。
10	C	**【题干】**：Krauser 将她继承的遗产 $6,000 用来投资，其中一部分的年单利为 9%，剩余部分的年单利为 12%。如果她一年所获得的总利息为 $660。 **【问题】**：她的投资中年单利为 12% 的那部分有多少？ **【解析】**：根据题意，设年单利为 12% 的那部分有 t，因此剩余部分为 $6,000-t$。可得以下关系式：$12\%\times x+9\%\times(6,000-x)=660$，解得 $x=4,000$，选 C。
11	A	**【题干】**：Barbara 和 Marc 两人分别投掷骰子 50 次，得到点数从 1 到 6 的频次如表所示。现在如果将两个人的统计频次汇总到一起。 **【问题】**：一共投掷骰子 100 次所得到的点数其中位数是多少？ **【解析】**：根据题意可得，两个人合在一起一共投掷骰子 100 次，中位数值应该是第 50 次和第 51 次点数的算术平均值。从表中可以看出，小于点数 3（点数 1 和点数 2）一共有 11（点数 1）+21（点数 2）=32 次，而点数为 3 的有 21 次，可见，从第 33 次一直到第 53（32+21=53）次之前，点数都为 3，因此第 50 次和第 51 次点数的算术平均值也为 3。选 A。

12	D	【题干与问题】：已知 $[(2x+y)+(x+2y)]^2$，下面哪一个表达式与此式子相等？ 【解析】：根据题意可得 $[(2x+y)+(x+2y)]^2 = (3x+3y)^2 = [3(x+y)]^2 = 9(x+y)^2$，选 D。
13	D	【题干】：为了去参加一次商务会议，Joanna 用了 h 小时开车 m 英里，结果提前半个小时到了。 【问题】：如果他要准时到达，需要行驶的车速是多少？ 【解析】：根据题意可得，如果要准时达到，需要 m 英里的路程花费时间为 $h+\frac{1}{2}$。因此需要行驶的车速为 $\frac{m}{h+\frac{1}{2}} = \frac{2m}{2h+1}$，选 D。
14	C	【题干】：两个圆柱形的水缸高度一致，其中较大水缸的半径等于较小水缸的直径，如果较大水缸的体积比较小水缸多 $k\%$。 【问题】：k 的值为多少？ 【解析】：根据题意，设较小水缸的半径为 r，高为 h，较大水缸的半径即为 $2r$，高为 h。可得 $V_{大} = \pi \times (2r)^2 \times h = 4\pi r^2 h$，$V_{小} = \pi \times (r)^2 \times h = \pi r^2 h$，可见较大水缸的体积是较小水缸体积的四倍，即增加了三倍。所以 $k = 300$，选 C。
15	A	【题干与问题】：下面两个方程式的图像有几个交点？ 【解析】：根据题意可得，两个方程式的图像位置如图 5-6 所示： 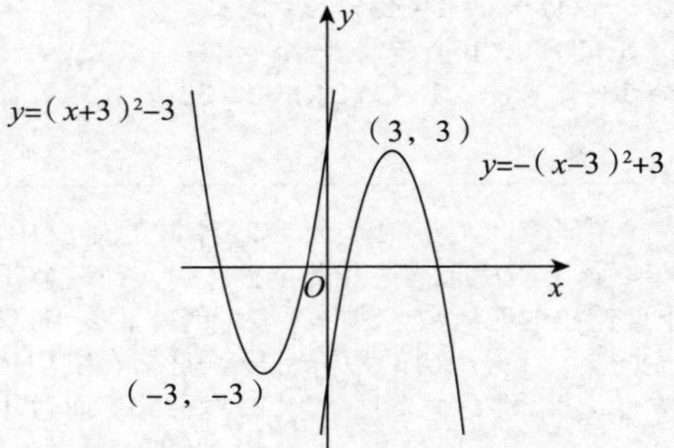 图 5-6 可见两个图像没有交点，选 A。
16	1	【题干与问题】：已知 $i = \sqrt{-1}$，且 $(1+i) \div (1-i) = a+bi$，如果 a 和 b 都是实数，则 $a+b$ 的值为多少？ 【解析】：根据题意，将原式变形为： $\frac{1+i}{1-i} = \frac{(1+i)(1+i)}{(1-i)(1+i)} = \frac{2i}{2} = i$，可见 $a=0$，$b=1$，因此 $a+b=1$。
17	1.5 or $\frac{3}{2}$	【题干与问题】：表中给出的是函数 $h(x) = \frac{x^2}{2} + k$ 相对应的 x 和 $h(x)$ 的值的集合，则 k 的值为多少？ 【解析】：根据题意，将 $x=3$，$h(x)=6$ 代入到函数中得到 $6 = \frac{3^2}{2} + k$，得到 $k=1.5$ 或者 $\frac{3}{2}$。

18	$\dfrac{7}{3}$	【题干与问题】：如图所示，$AB=BD=DF=FH=1$，则梯形 $FHJG$ 与 $BCED$ 的面积之比为多少？ 【解析】：从给出的图中可以看出△ABC、△ADE、△AFG 和△AHJ 都是以∠A 为顶角的相似三角形，再根据原题所给的条件可以得到线段 $AB:AD:AF:AH=1:2:3:4$。同理还可以得到线段 $BC:DE:FG:HJ=1:2:3:4$。设线段 BC 的长度为 a，因此梯形 $FHJG$ 的面积为 $\dfrac{1}{2}\times(3a+4a)\times 1=\dfrac{7a}{2}$，梯形 $BCED$ 的面积为 $\dfrac{1}{2}\times(a+2a)\times 1=\dfrac{3a}{2}$，所以梯形 $FHJG$ 与 $BCED$ 的面积之比为 $\dfrac{7}{3}$。
19	2	【题干】：如图所示，$ABCD$ 是一个正方形，边长为 2，点 E 是边 AB 的中点，点 F 是边 AD 的中点。 【问题】：四边形 $CFAE$ 的面积是多少？ 【解析】：根据题意可得，四边形 $CFAE$ 的面积可以用：正方形 $ABCD$ 的面积 $-$ △FCD 的面积 $-$ △BCE 的面积得到。从原题给出的条件可以得到线段 $CD=$ 线段 $BC=2$，线段 $DF=$ 线段 $BE=1$，因此△FCD 的面积 $=$ △BCE 的面积 $=\dfrac{1}{2}\times 2\times 1=1$，因此四边形 $CFAE$ 的面积 $=4-1-1=2$。
20	1	【题干与问题】：如果直线 $x+2y=7$ 和直线 $2x-ky=5$ 垂直，则 k 的值为多少？ 【解析】：根据题意，将两条直线分别写成标准方程的形式： 直线 1：$y=\dfrac{7}{2}-\dfrac{x}{2}$，直线 2：$y=\dfrac{2}{k}x-\dfrac{5}{k}$，如果两条直线垂直，则它们的斜率互为负倒数，即 $\dfrac{2}{k}=2$，解得 $k=1$。

Section 4

1	D	【题干与问题】：如果 $\lvert 10-3y\rvert<3$，下面哪一个值可能是 y 的值？ 【解析】：根据题意可得 $10-3y<3$ 或者 $10-3y>-3$，解得 $y>\dfrac{7}{3}$，满足条件的只有 D 项。
2	D	【题干】：三辆车通过高速公路的一个限速标志，A 车的速度是 B 车的两倍，C 车的速度比 B 车快 20 英里/小时，如果 C 车的速度是 60 英里/小时。 【问题】：A 车的速度是多少？ 【解析】：根据题意可得，B 车的速度是 $60-20=40$ 英里/小时，A 车的速度是 $40\times 2=80$ 英里/小时，选 D。
3	B	【题干与问题】：如果满足 $-1<x<0$，那么下列哪一个式子一定正确？ 【解析】：根据题意，此题可以设相应的 x 值来验证。设 $x=-\dfrac{1}{2}$，因此 $\dfrac{x}{2}=-\dfrac{1}{4}$，$x^2=\dfrac{1}{4}$，$x^3=-\dfrac{1}{8}$，可得Ⅰ错误，Ⅱ正确，Ⅲ错误，选 B。
4	B	【题干】：一位富人的财产分配如下：财产的 10% 分给他的妻子，财产的 5% 平均分给他的三个孩子，财产的 5% 平均分给他的五个孙子，剩余的财产交付给一个慈善信托公司托管，如果该慈善信托公司获得 \$1,000,000。 【问题】：该富人的每一位孙子继承了多少钱？ 【解析】：根据题意可得，该富人将全部财产的 $100\%-10\%-5\%-5\%=80\%$ 给了该慈善信托公司托管，设他的总财产为 T，则 $T\times 80\%=1,000,000$，可得 $T=1,250,000$。他的孙子每个人继承的钱 $=\dfrac{1,250,000\times 5\%}{5}=12,500$，选 B。

5	C	【题干】：该柱状图表示的是四所大学 2015 年向学校申请助学金的学生数量。四所大学对每个学生的平均资助额度分别为 $15,500、$21,000、$18,700 和 $14,300。 【问题】：哪个大学拨付给学生的助学金总数最多？ 【解析】：根据题意，A 大学共有 4,000 名学生获得资助，助学金总数为 $15,500 × 4,000 = $62,000,000，B 大学共有 2,800 名学生获得资助，助学金总数为 $21,000 × 2,800 = $58,800,000，C 大学共有 3,600 名学生获得资助，助学金总数为 $18,700 × 3,600 = $67,320,000，D 大学共有 4,500 名学生获得资助，助学金总数为 $14,300 × 4,500 = $64,350,000，可见 C 大学的总数最多，选 C。
6	A	【题干】：Marie 拥有一个网站售卖 CD 和 DVD，她进货价格 CD 每张 $2.75，DVD 每张 $5.75。如果需要邮寄给顾客，每张 CD 或者 DVD Marie 需要支付 $0.95 邮费。现在她对每张 CD 定价 $4.99，每张 DVD 定价 $9.99，并且对每张 CD 或者 DVD 加收邮费和手续费为 $1.75。 【问题】：下面哪一个式子表示的是她卖了 x 张 CD 和 y 张 DVD 后的利润 P 美元？ 【解析】：根据题意，利润＝收入－成本，因此每张 CD 的利润为 $4.99 + $1.75 − $2.75 − $0.95 = $3.04，每张 DVD 的利润为 $9.99 + $1.75 − $5.75 − $0.95 = $5.04。所以总的利润为 $P = 3.04 \times x + 5.04 \times y$，选 A。
7	C	【题干】：在某一州有 500 位登记选民，分为民主党人、共和党人和独立人士。一项调查统计了在一次对"提案 8"的公民表决提案中选民各自的投票情况。统计结果如表中所示。在投票当天，所有支持该提案的选民都投了赞成票，所有不支持该提案的选民都投了反对票。此外，在未决定投票的选民中有 80% 的人支持这个提案，20% 的人反对。 【问题】：在调查的 500 人中，支持该提案的人所占的百分比是多少？ 【解析】：根据题意可得，支持该提案的人分为 2 批： 参加投票的选民中有 192 人支持，未决定投票的选民中有 90 × 80% = 72 人支持，共计有 192 + 72 = 264 人，占总数 $\frac{264}{500} \times 100\% = 52.8\%$。选 C。
8	C	【题干】：根据美国人口统计局的统计，在美国每 8 秒诞生一个婴儿，每 12 秒有一人去世，此外基于人口的迁出和迁入，每 30 秒有一个人的净增加。 【问题】：美国人口的日平均增加数与下面哪一个值最接近？ 【解析】：根据题意，原题中有不同的时间段，因此首先需要统一，即找到 8、12 和 30 的最小公倍数，为 120。因此每 120 秒有 15 个婴儿诞生、有 10 个人去世，有 4 个人净增加，总计增加 15 − 10 + 4 = 9 人，1 天有 24 × 60 × 60 = 86,400 秒，因此日平均增加数为 $9 \times \frac{86,400}{120} = 6,480$。选 C。
9	C	【题干】：在 2015 年 1 月，美国的人口达到 320,000,000。 【问题】：根据人口统计局的分析，到哪一年该国的人口可以达到 350,000,000？ 【解析】：根据上一题可得美国年平均人口增长数量为 6,480 × 365 = 2,365,200。因此需要 $\frac{350,000,000 - 320,000,000}{2,365,200} \approx 13$ 年，即 2015 + 13 = 2028 年，选 C。

10	D	【题干】：如图所示，一个圆锥形水罐的底面半径为 3 英尺，高为 6 英尺，如果水罐充满水后正好倒出一半体积的水。 【问题】：水罐中剩余水的高度是多少英寸？ 【解析】：根据题意可得，水罐倒出一半水后，剩余的部分也是一个圆锥形，且这两个圆锥形相似，如图 5-7 所示。设剩余部分所形成的圆锥形底面半径为 r，高为 h。原圆锥形的底边半径与高之比为 $1:2$，因此，与它相似的小圆锥形也符合这个比例，即 $r=\frac{1}{2}h$。 水罐全部充满水的体积为 $\frac{1}{3}\times\pi\times 3^2\times 6=18\pi$，倒掉一半水之后的体积为 $9\pi=\frac{1}{3}\times\pi\times r^2\times h=\frac{1}{12}\times\pi\times h^3$，解得 $h\approx 4.76$ 英尺≈ 57 英寸。选 D。 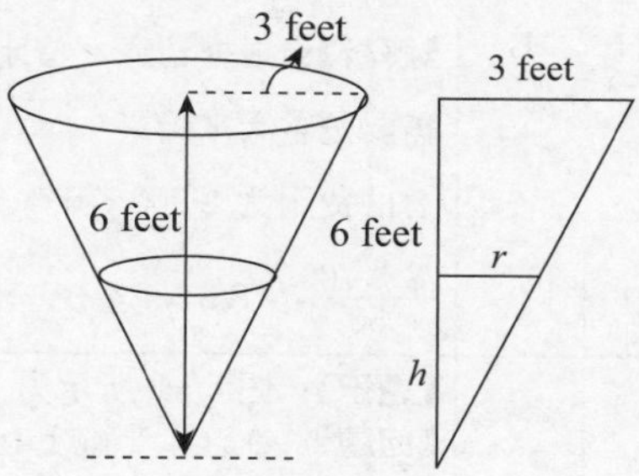 图 5-7
11	D	【题干】：2000 年，Jennifer 投资 \$1,000 到一个为期 7 年的存款单，该存款的年复利息为 2%。到了 2007 年，该存款到期了，她将所有的钱又投资了另一个为期 7 年的存款单，也是 2% 的年复利息，到 2014 年到期。 【问题】：从 2007 年到 2014 年这 7 年她得到的钱比从 2000 年到 2007 年这 7 年得到的钱多多少？ 【解析】：根据题意，由于是复利，所以到了 2007 年她所得到的本金加上利息一共是 $(1.02)^7\times\$1,000=\$1,148.69$。她又将这部分钱投入到另一个存单，到 2014 年，她所得到的本金加上利息一共是 $(1.02)^7\times\$1,148.69=\$1,319.48$，所以第二个 7 年期比第一个 7 年期多 $\$1,319.48-\$1,148.69=\$170.79\approx\$171$，选 D。
12	D	【题干与问题】：如果函数 f 为 $f(x)=x^2+ax+a$，a 是一个常数，则 $f(5)$ 用 a 的形式来表示是多少？ 【解析】：根据题意可得 $f(5)=5^2+5a+a=25+6a$，选 D。
13	B	【题干与问题】：下面方程组解的集合是哪一个？ 【解析】：根据题意将 $y=3+x$ 代入第二个方程中得到：$x^2-7(3+x)+31=0$，化简得 $(x-5)(x-2)=0$，解得 $x=5$，$y=8$，或者 $x=2$，$y=5$。因此方程组解的集合为 $(2,5)$ 和 $(5,8)$，选 B。
14	B	【题干与问题】：求满足不等式 $\frac{x}{12}-\frac{x+2}{4}<0$ 的 x 值。 【解析】：根据题意，原不等式可以变形为： $\frac{x-(3x+6)}{12}<0$，化简为 $-2x-6<0$，解得 $x>-3$，选 B。
15	B	【题干与问题】：如果 $\cos\frac{\pi}{3}=x-1$，则 x 的值是多少？ 【解析】：根据题意可得 $\cos\frac{\pi}{3}=\frac{1}{2}=x-1$，解得 $x=\frac{3}{2}$，选 B。
16	A	【题干与问题】：如果 $x\neq 0$，$y\neq 0$ 且 $x\neq y$，下面哪一个式子与 $\frac{x^{-1}-y^{-1}}{x-y}$ 一致？ 【解析】：根据题意可得 $\frac{x^{-1}-y^{-1}}{x-y}=\frac{\frac{1}{x}-\frac{1}{y}}{x-y}=\frac{\frac{y-x}{xy}}{x-y}=\frac{-(x-y)}{xy(x-y)}=-\frac{1}{xy}$，选 A。

17	B	**【题干】**：一个乐队录制一个唱片，他们租赁了一间工作室，每天的费用为 $200，可以使用 12 小时。每天的费用不可以再按比例分配，因此工作室的租赁时间必须按整天来购买。他们的音响设备每小时的花费为 $28，音响的使用时间为整个录制时间的一半，如果完成整个录制需要 50 小时。 **【问题】**：整个录制期间平均每小时的成本是多少？ **【解析】**：根据题意，总计需要使用工作室的天数为：$\frac{50}{12} = 4.2$ 天 ≈ 5 天（原题条件中指出需要按整天来购买）。 总的费用为 $200 \times 5 + 28 \times \frac{50}{2} = 1,700$，因此整个录制期间平均每小时的成本为：$\frac{1,700}{50} = 34$，选 B。
18	B	**【题干】**：图 5-8 所示为一个半圆，圆心在 (0, 0)。 **【问题】**：在这个半圆上有两个点其 x 轴的坐标值相等，求这两个点的 y 轴坐标值是多少？ **【解析】**：根据题意可得，在该半圆中，这两个点的 x 轴坐标对应的值相等，y 轴坐标值必然为相反数，如图所示： 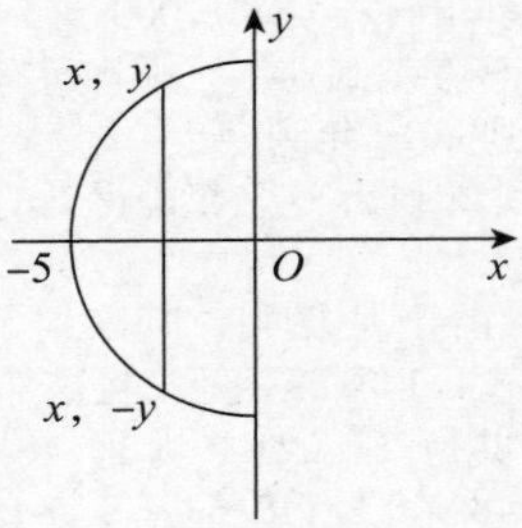 图 5-8 比较四个选项发现符合该条件的只有 B 项，选 B。
19	B	**【题干】**：一个数列各数字间的差值（即公差）是偶数，第 1 项是 7，第 20 项是 159。 **【问题】**：该数列的第 4 项是下面的哪一个数字？ **【解析】**：根据题意可得，159 与 7 之间一共还有 19 个数字，因此可得公差为 $\frac{159-7}{19} = 8$。 因此第 4 项为 $a_4 = a_1 + (n-1) \times d = 7 + 3 \times 8 = 31$，选 B。
20	C	**【题干】**：一个班有 100 名学生，其中 65 名学生选修西班牙语，32 名学生选修艺术，14 名学生两个科目都选。 **【问题】**：有多少学生两个科目都不选？ **【解析】**：根据题意可得一个维恩图，如图 5-9： 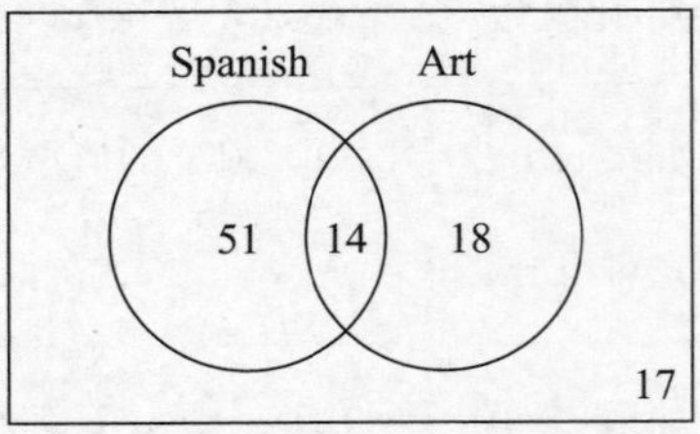 图 5-9 首先计算 $65 + 32 = 97$，扣除 14 名学生两个科目都选，剩下 $97 - 14 = 83$ 名学生选择其中一个，最终剩余 $100 - 83 = 17$ 名学生两科都不选，选 C。

21	B	【题干与问题】：如果表达式 $\frac{5-9x^2}{2-3x}$ 可以写成 $\frac{1}{2-3x}+A$ 的形式，那么用 x 来表示 A 为多少？ 【解析】：根据题意可得 $\frac{1}{2-3x}+A=\frac{1}{2-3x}+\frac{2-3x}{2-3x}\times A=\frac{1+2A-3xA}{2-3x}=\frac{5-9x^2}{2-3x}$，因此得到 $A\times(2-3x)=4-9x^2$，因此 $A=\frac{4-9x^2}{2-3x}=\frac{(2+3x)(2-3x)}{2-3x}=2+3x$，选 B。
22	D	【题干】：在如下的数列中，首项为 2，从第 2 项开始，每项都比前面一项的 3 倍少 3。 【问题】：k 的值为多少？ 【解析】：根据题意可得，k 应该比 6 的 3 倍少 3，因此 k 为 $6\times 3-3=15$，选 D。
23	C	【题干】：如图给出了函数 f 和函数 g 的图像，其中函数 f 定义为 $f(x)=2|x+2|$，函数 g 定义为 $g(x)=f(x+h)+k$，且 h 和 k 都是常数。 【问题】：$|h-k|$ 的值为多少？ 【解析】：根据题意可得函数 g 的图像是由函数 f 的图像向右移动 3 个单位，向上移动 2 个单位所得到的。因此可以得到 $g(x)=f(x-3)+2$，与原题给出的条件对应后可以得到 $h=-3$，$k=2$。因此 $|h-k|=|-3-2|=5$，选 C。
24	C	【题干】：设 a，b 和 c 的算术平均值为 h，b，c 和 d 的算术平均值为 j，d 和 e 的平均值为 k。 【问题】：a 和 e 的平均值是多少？ 【解析】：根据题意可得，$a+b+c=3h$，$b+c+d=3j$，$d+e=2k$。将三个式子合并可得 $a+e=3h-3j+2k$，所以 a 和 e 的平均值为 $\frac{3h-3j+2k}{2}$，选 C。
25	A	【题干】：John 用票来支付费用给他的工人，每张票 \$32。每个工人每小时挣 \$60，John 一共有 24 张票。 【问题】：下面哪一个函数表示的是开始工作 t 小时后还剩的票的张数？ 【解析】：根据题意可得，每个工人每小时可以挣得 $\frac{60}{32}$ 张票，工作 t 小时，一共得到的票数为 $\frac{60t}{32}$ 张，因此剩余 $24-\frac{60t}{32}$ 张，选 A。
26	D	【题干】：一位研究者对估算妈妈带着 4 岁的孩子在下午 5 点到 10 点看电视的平均时间比较感兴趣。她调查了 80 位妈妈计算出平均看电视的时间为 34 分钟，误差范围为 3.1 分钟。如果她想重复这个调查以期望得到更小的误差范围。 【问题】：下面哪一种采样方式可以减少这个误差范围？ 【解析】：根据题意，要想缩小误差范围，需要扩大满足调查条件的样本，因此在本题中需要更多符合调查条件的妈妈，因此选 D。
27	C	【题干】：在 Corbett 小学去年有四个不同的班级参加了两个筹款活动，一个是在二月，一个是在五月，每个班的参与率如图所示。 【问题】：哪个班在两次筹款活动中参与率的变化最大？ 【解析】：根据题意可得，A 班二月筹款活动参与率为 60%，五月筹款活动参与率为 60%，B 班两次参与率分别为 80% 和 70%，C 班两次参与率分别为 50% 和 70%，D 班两次参与率分别为 90% 和 90%。可见 C 班参与率变化最大，选 C。

28	C	【题干】：如果 A 和 C 班各有 20 名学生，B 和 D 班各有 30 名学生。 【问题】：一共有多少学生参加了五月的筹款活动？ 【解析】：根据题意可得：A 班和 C 班五月筹款活动参与率分别为 60% 和 70%，有 $20 \times 60\% + 20 \times 70\% = 12 + 14 = 26$ 人参加，B 班和 D 班五月筹款活动参与率分别为 70% 和 90%，有 $30 \times 70\% + 30 \times 90\% = 21 + 27 = 48$ 人参加，因此共有 $26 + 48 = 74$ 人参加，选 C。
29	C	【题干与问题】：如该方程式所示，c 是常数。如果 $x = -1$ 是该方程式的一个解。 【问题】：该方程式的另一个解为多少？ 【解析】：根据题意将 $x = -1$ 代入到原式当中去可得 $3 \times (-1)^2 = 4 \times (-1) + c$，得到 $c = 7$。因此原式为 $3x^2 - 4x - 7 = 0$，转化为 $(x+1)(3x-7) = 0$，除了 $x = -1$ 之外，还有一个解为 $x = \dfrac{7}{3}$，选 C。
30	D	【题干】：在你寻找暑期打工的时候，你可以获得以下的机会。机会 1：在 Timmy 玉米卷店，你每小时可以挣得 \$4.50，但是需要购买价格为 \$45.00 的制服。每周你可以计划工作 20 小时。机会 2：在 Kelly 洗车行，你每小时可以挣得 \$3.50，没有其他的要求，但是你必须同意每周工作 20 小时。 【问题】：在做决定之前，你可以考虑所有的因素，下面哪一项的结论不正确？ 【解析】：根据题意可得，在 Kelly 洗车行工作 8 周，可以获得 $\$3.50 \times 20 \times 8 = \$560$，A 项正确。在 Timmy 玉米卷店工作 10 小时，可获得 $\$4.50 \times 10 = \$45$，正好抵消购买制服所需要的费用，B 项正确。工作 2 周，在 Timmy 玉米卷店的收入为 $\$4.50 \times 20 \times 2 - \$45 = \$135$，而在 Kelly 洗车行的收入为 $\$3.50 \times 20 \times 2 = \$140$，相比之下 Kelly 洗车行收入更高，C 项正确。假设工作了 41 小时，在 Timmy 玉米卷店的收入为 $\$4.50 \times 41 - \$45 = \$139.5$，而在 Kelly 洗车行的收入为 $\$3.50 \times 41 = \$143.5$，可见 Kelly 洗车行支付更多，D 项不正确，选 D。
31	50	【题干】：Lauren 的储蓄账户中有 \$80，当她拿到薪水后，她存了一些钱使得这个账户中的钱为 \$120。 【问题】：她的账户中钱增加的百分比是多少？ 【解析】：根据题意可得，总存款为 \$120，增加了 $\$120 - \$80 = \$40$，增加的百分比为 $\dfrac{40}{80} \times 100\% = 50\%$。
32	0.84	【题干与问题】：如果 $\cos(x - \pi) = 0.4$，则 $\sin^2 x$ 的值为多少？ 【解析】：如图 5-10，根据诱导公式可得：$\cos x = -\cos(x - \pi) = -0.4$，因此 $\sin^2 x = 1 - \cos^2 x = 1 - (-0.4)^2 = 0.84$。 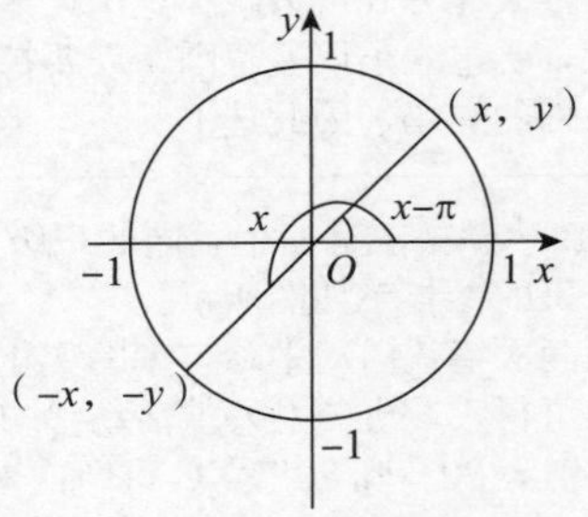 图 5-10

33	0.05	【题干】：一位分析家希望用该公式估算一个初始投资为 $1,000 的互助基金经过 m 个季度之后的金额。如果两个季度后初始投资为 $1,000 的互助基金金额变成了 $1,102.50。 【问题】：公式中 k 的值为多少？ 【解析】：根据题意，将原题给出的条件代入公式中：$1,102.50 = 1,000 \times (1+k)^2$，得到 $1+k = 1.05$，解得 $k = 0.05$。
34	384	【题干与问题】：如果直线 l 通过原点，并且与直线 $y = \frac{2}{3}x - 6$ 平行，直线 l 与直线 $y = \frac{1}{2}x - 4$ 相交于点 (x, y)，则 xy 的乘积是多少？ 【解析】：根据题意，直线 l 与直线 $y = \frac{2}{3}x - 6$ 平行，因此斜率也为 $\frac{2}{3}$，直线 l 过原点，所以在 y 轴上的截距为 0，这样可得直线 l 的方程为 $y = \frac{2}{3}x$。直线 l 与直线 $y = \frac{1}{2}x - 4$ 相交于点 (x, y)，所以可以将这两个方程联立得到方程组，求得的 x 和 y 值即是交点的坐标：$$\begin{cases} y = \frac{2}{3}x \\ y = \frac{1}{2}x - 4 \end{cases}$$ 解得 $x = -24$，$y = -16$，x 和 y 的乘积为 384。
35	52	【题干与问题】：给出的方程式是 xy 平面坐标系中直线的方程，h 是一个常数，如果这条直线的斜率为 -13，则 h 的值为多少？ 【解析】：根据题意，将直线方程写成标准形式为 $y = -\frac{h}{4}x - \frac{3}{4}$，斜率为 $-\frac{h}{4} = -13$，解得 $h = 52$。
36	24	【题干】：如图所示，块状物体底边是一个六边形，该六边形的边长和内角都分别相等，每一个侧面都是长方形。 【问题】：该块状物的一个侧面的面积是多少平方英尺？ 【解析】：根据题意，该六边形的边长和内角都分别相等，说明是一个正六边形，每一个内角都相等，为 $120°$。在该正六边形内的六个内部的三角形均为等边三角形，且边长为六边形对角线长度的一半，为 12 英寸。因此一个侧面的面积为 $2 \times 12 = 24$ 平方英寸。
37	71	【总题干】：一家制鞋公司生产多个批次的皮鞋和靴子，每一批次都用 2,000 平方英尺的皮革进行生产。一双靴子需要 10 平方英尺的皮革，一双皮鞋需要 3 平方英尺的皮革。 【题干】：该公司需要生产一个批次的鞋子 500 双。 【问题】：该批次中靴子为多少双？ 【解析】：根据题意，设该批次中皮鞋数量为 s 双，靴子为 b 双，可得一个方程组：$$s + b = 500$$ $$3s + 10b = 2,000$$ 解得 $b = 71.4$，注意由于是实物，需要逻辑取舍，因此 $b = 71$ 或者 72，再代入上述方程组验证：如果 $b = 71$，$s = 429$，则 $3 \times 429 + 10 \times 71 = 1,997 < 2,000$，满足条件，如果 $b = 72$，$s = 428$，则 $3 \times 428 + 10 \times 72 = 2,004 > 2,000$，不满足条件，所以 $b = 71$。

38	160	【题干】：一位顾问参与评估该公司的生产过程，他建议基于市场分析的情况来评价。由于对靴子的需求增加了，一双靴子可以卖到$300，而一双皮鞋可以卖$75，假设其他生产成本都一样。 【问题】：该公司需要卖出多少双靴子使得其利润最大？ 【解析】：根据题意可得，两种鞋类之间的关联就是使用皮革作为原材料，因此首先需要折算出对应于相同原材料的不同价格： 靴子的单位原材料价格为 $\frac{300}{10} = \$30$，皮鞋的单位原材料价格为 $\frac{75}{3} = \$25$，因此可得 $\frac{b}{s} = \frac{6}{5}$，可以联立得到一个方程组： $$\begin{cases} 3s + 10b = 2,000 \\ \frac{b}{s} = \frac{6}{5} \end{cases}$$ 得到 $b = 160$。

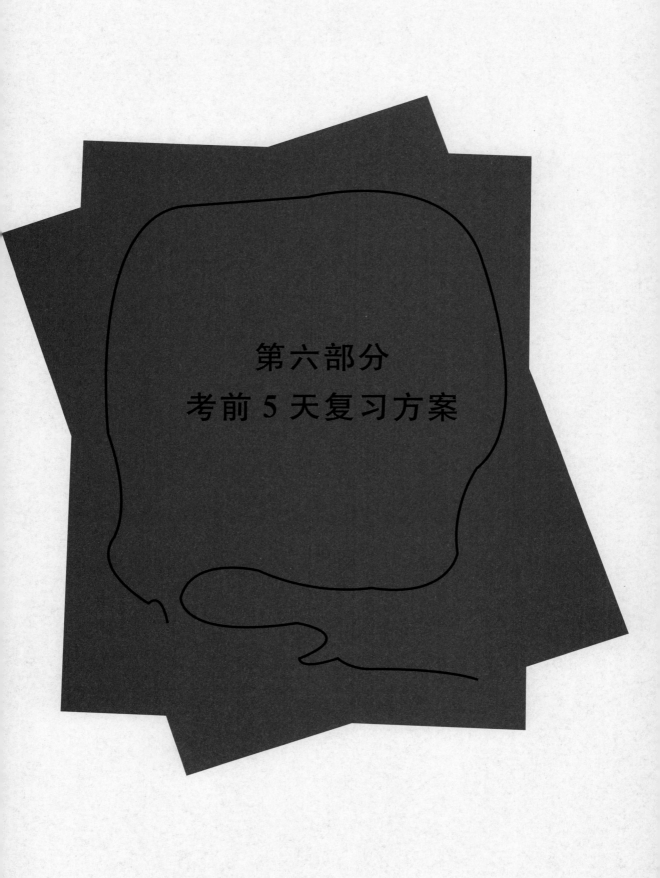

第六部分
考前 5 天复习方案

Day 1

Day 1 复习方案

复习内容

☐ 数与数的运算

☐ 因子与倍数

☐ 比例

☐ 百分比

考前适应性练习

In the figure (Fig. 6-1), the average acreage per farm was approximately 140 in 1910 and 220 in 1950. The ratio of the total farmland acreage in 1910 to the total in 1950 was most nearly:

A. $\frac{3}{4}$ B. $\frac{2}{3}$ C. $\frac{3}{5}$ D. $\frac{1}{2}$

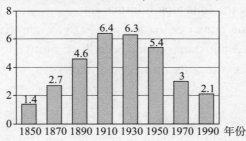

Fig. 6-1

If at the end of 1973 Company X sold 30,000 shares of common stock for 35 times Company X's earning for the year (Fig. 6-2), what was the price of a share of common stock at that time?

A. $7.00   B. $10.00   C. $17.50   D. $35.00

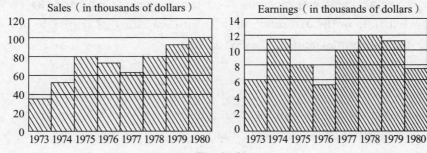

Fig. 6-2

There were 1,500 textbooks in School X. In general, the number of the textbooks into the inventory was 10% of total textbooks in a given year. If School X purchased 300 textbooks in 1971 and all of these textbooks either were counted in the inventory or had been discarded before the inventory, what percent of these textbooks had been discarded?
A. 10%　　　　B. 20%　　　　C. 50%　　　　D. 80%

The table (Table 6-1) shows the average (arithmetic mean) price per dozen of the large grade A eggs sold in a certain store during three successive months. If $\frac{2}{3}$ as many dozen were sold in April as in May, and twice as many were sold in June as in April, what was the average price per dozen of the eggs sold over the three-month period?
A. $1.08　　　　B. $1.10　　　　C. $1.14　　　　D. $1.16

Table 6-1

Month	Average Price per Dozen
April	$1.26
May	$1.20
June	$1.08

Questions 5 and 6 refer to the following information.
Three cars all arrive at the same destination at 4:00 PM. The first car traveled 144 miles mostly by highway. The second car traveled 85 miles mainly on rural two-lane roads. The third car traveled 25 miles primarily on busy city streets.

The first car traveled at an average speed of 64 mph. The second car started its drive at 2:18 PM. How many minutes had the first car already been traveling before the second car started its drive?

The third car encountered heavy traffic for the first 60% of its trip and only averaged 15 mph. Then traffic stopped due to an accident, and the car did not move for 20 minutes. After the accident was cleared, the car averaged 30 mph for the remainder of the trip. At what time in the afternoon did the third car start its trip? Use only digits for your answer (For example, enter 1:25 PM as 125.).

Max purchased some shares of stock at $10 per share. Six months later the stock was worth $20 per share. What was the percent increase in the value of Max's investment?
A. 20%　　　　B. 50%　　　　C. 100%　　　　D. 200%

On January 1, 2015, the values of Alice's brokerage account and of Barbara's brokerage account were a dollars and b dollars, respectively. During the year, the value of Alice's account increased by 10% and the value of Barbara's account decreased by 10%. If on December 31, 2015 the values of their accounts were equal, what is the ratio of a to b?

On Thursday, 20 of the 25 students in a chemistry class took a test, and their average (arithmetic mean) was 80. On Friday, the other 5 students took the test, and their average (arithmetic mean) was 90. What was the average for the entire class?
A. 82　　　　　　B. 84　　　　　　C. 85　　　　　　D. 88

Susan is a candidate for mayor of a city that has 20 election precincts. She assigned each of her 84 volunteers to work in one of the precincts. Each precinct that has 1,000 or more registered voters has been assigned 5 volunteers, and each precinct that has fewer than 1,000 registered voters has been assigned 3 volunteers. What percent of the precincts have fewer than 1,000 registered voters?
A. 25%　　　　　　B. 40%　　　　　　C. 60%　　　　　　D. 75%

■答案与解析

1	A	【题干】：图中给出了1850—1990年美国农场的数量，在1910年每个农场的面积大约是140英亩，在1950年每个农场的面积大约是220英亩。 【问题】：1910年农场的总面积与1950年农场的总面积之比最接近下面哪一个值？ 【解析】：从图上可知，1910年农场总数为6.4百万，1950年农场总数为5.4百万，因此两者面积之比等于：$\frac{6.4 \times 140}{5.4 \times 220} = 0.752 \approx 0.75$，选A。
2	A	【题干】：假如X公司在1973年底以该公司一年盈利的35倍价钱售出30,000份普通股。 【问题】：当时普通股的价钱是多少？ 【解析】：题中sell for指买进或卖出的价钱。根据图可得到X公司在1973年的盈利是$6,000，所以每股普通股的价钱为：$\frac{35 \times \$6,000}{30,000} = \$7.00$，选A。
3	C	【题干】：X学校共有1,500本课本。一般来说，当年入库的书的数量要占总书的数量的10%。假设X学校在1971年购买了300本课本，这些书要么入库要么在入库前被废弃。 【问题】：被废弃的课本占这些书的百分之几？ 【解析】：在1971年，X学校入库图书数目为：$1,500 \times 10\% = 150$，所以被废弃课本所占的百分比为：$\frac{300 - 150}{300} \times 100\% = 50\%$，选C。
4	D	【题干】：表格中表明了某商店连续三个月的大A级鸡蛋每打的平均售价。若4月份销售鸡蛋的打数是5月份的$\frac{2}{3}$，且6月份销售鸡蛋的打数是4月份的2倍。 【问题】：这三个月期间每打鸡蛋的平均售价是多少？ 【解析】：设4月份的销售打数为x，则5月份销售的打数为$\frac{3}{2}x$，6月份销售的打数为$2x$，可

		得这三个月每打鸡蛋的平均销售价格为：$\dfrac{\$1.26x + \$1.2 \times \dfrac{3}{2}x + \$1.08 \times 2x}{x + \dfrac{3}{2}x + 2x} = \$1.16$，选 D。
5	33	【题干】：三辆车在下午 4 点同时到达某一目的地。第一辆车走高速公路，行驶了 114 英里，第二辆车走农村双车道公路，行驶了 85 英里，第三辆车走城市繁忙街道，行驶了 25 英里。第一辆的速度为 64 英里/小时，第二辆车是在下午 2:18 出发的。 【问题】：在第二辆车出发之前第一辆车已经行驶了多少时间？ 【解析】：根据题意可得，第一辆车走了 144 英里，设第一辆行驶的总时间为 t，$64 \times t = 144$，解得 $t = 2.25$ 小时 = 135 分钟。第二辆车在 2:18 出发，4:00 到，行驶了 1 小时 42 分钟即 102 分钟。因此在第二辆车出发之前第一辆车已经行驶的时间为 135 − 102 = 33 分钟。
6	220	【题干】：在行驶到总路程 60% 的时候第三辆车遇到了交通拥挤，因此速度只有 15 英里/小时。之后由于一个事故，这辆车停了 20 分钟。在这个事故之后，该辆车以 30 英里/小时的速度行驶完余下的路程。 【问题】：第三辆车在下午什么时候出发的？（仅用数字表示答案，比如 1:25 写成 125） 【解析】：根据题意可得，第三辆车已经行驶了 60%，因此行驶了 25×60% = 15 英里，设在这段距离第三辆车行驶的时间为 t_1，可以得到 $15 \times t_1 = 15$，解得 $t_1 = 1$ 小时 = 60 分钟。该车剩余的路程为 25×40% = 10 英里，设第三辆剩余行驶时间为 t_2，因此可以得到 $30 \times t_2 = 10$，$t_2 = 20$ 分钟。这样可以得到第三辆所用的总时间为 60 + 20 + 20 = 100 分钟 = 1 小时 40 分钟，这样从下午 4:00 向前推 1 小时 40 分钟得到 2:20 PM，因此结果为 220
7	C	【题干】：Max 购买了一股票，每股的价格为 \$10，6 个月之后，该股票价值为 \$20。 【问题】：Max 投资量增加的百分比为多少？ 【解析】：根据题意可得，该股票一开始为 \$10，6 个月增加到 \$20，增加了 \$10，因此增加的百分比为 $\dfrac{20-10}{10} \times 100\% = 100\%$，选 C。
8	$\dfrac{9}{11}$	【题干】：2015 年 1 月 1 日，Alice 的经纪人账户价值为 a 美元，Barbara 的经纪人账户价值为 b 美元。在这一年，Alice 的账户价值增加了 10%，而 Barbara 的账户价值下降了 10%。在 2015 年 12 月 31 日，他们两个的账户价值相等了。 【问题】：a 与 b 比值为多少？ 【解析】：根据题意可得，在 2015 年 12 月 31 日，Alice 的账户价值为 $a + 10\% \times a = a + 0.1a = 1.1a$，Barbara 的账户价值为 $b - 10\% \times b = b - 0.1b = 0.9b$。可得 $1.1a = 0.9b$，因此 $\dfrac{a}{b} = \dfrac{0.9}{1.1} = \dfrac{9}{11}$。
9	A	【题干】：在周四的一次化学课上 25 名学生中的 20 名参加了一项测试，他们的平均成绩为 80 分。在周五，另外 5 名学生参加了这个测试，他们的平均成绩为 90 分。 【问题】：全班的平均成绩是多少？ 【解析】：根据题意可得，20 名学生的总成绩为 20×80 = 1,600 分，另外 5 名学生的总成绩为 5×90 = 450 分，全班总成绩为 1,600 + 450 = 2,050 分，因此全班的平均成绩是 $\dfrac{2,050}{25} = 82$ 分，选 A。
10	B	【题干】：Susan 是一个城市市长候选人，拥有 20 个选举选区。她将 84 名志愿者分配到每一个选区。如果一个选区拥有 1,000 或者更多的选民，则安排 5 位志愿者，如果该选区拥有的选民不足 1,000，则安排 3 位志愿者。 【问题】：拥有选民不足 1,000 的选区所占的比例是多少？ 【解析】：设拥有 1,000 或者更多选民的选区数量为 x，选民不足 1,000 的选区数量为 y，因此可得： $$x + y = 20$$ $$3x + 5y = 84$$ 解得 $x = 8$，$y = 12$，因此拥有选民不足 1,000 的选区所占的比例是 $\dfrac{8}{20} \times 100\% = 40\%$，选 B。

Day 2

Day 2 复习方案
复习内容
☐ 统计与概率
☐ 乘方与根数
☐ 代数
☐ 解析几何
考前适应性练习

Joseph joins a gym that charges \$79.99 per month plus tax for a premium membership. A tax of 6% is applied to the monthly fee. Joseph is also charged a one-time initiation fee of \$95 as soon as he joins. There is no contract so that Joseph can cancel at any time without having to pay a penalty. Which of the following represents Joseph's total charge, in dollars, if he keeps his membership for t months?
A. $1.06(79.99 + 95)t$
B. $1.06(79.99t + 95)$
C. $1.06 \times 79.99t + 95$
D. $(79.99 + 0.06t) + 95$

If $x + y = 5$ and $x^2 + 3xy + 2y^2 = 40$, find the value of $2x + 4y$.

Paris and Genevieve are waiting in line to buy tacos. It's Tuesday, and the taqueria has a Taco Tuesday special: all tacos are 50% off. Fish tacos normally sell for \$2.50 each, and beef and chicken tacos are normally \$1.50. They need exactly 16 tacos, and cannot spend more than \$15. What is the most they can spend on fish tacos?
A. \$6   B. \$7.5   C. \$10   D. \$15

If the expression $\dfrac{9x^2}{3x+5}$ is written in the equivalent form $\dfrac{25}{3x+5} + k$, what is k in terms of x?
A. $9x^2$
B. $9x^2 + 5$
C. $3x - 5$
D. $3x + 5$

A farmer randomly selected 40 ducks on the farm and measured their weight. The mean weight in the sample was 24 pounds, and the margin of error was 2.8 pounds. Farmer intends to replicate the survey and will attempt to get a smaller margin of error. Which of the following samples will most likely result in a smaller margin of error for the estimated mean weight of ducks?

A. 25 randomly selected ducks on the farm
B. 25 randomly selected animals on the farm
C. 100 randomly selected ducks on the farm
D. 100 randomly selected animals on the farm

If $(2^{32})^{(2^{32})} = 2^{(2^x)}$, what is the value of x?

The median annual salary of all the employees at Hartley's Home Supplies is \$45,000, whereas the range of their salaries is \$145,000. Which of the following is the most logical explanation for the large difference between the median and the range?

A. Half of the employees earn less than \$45,000
B. There is at least one employee who earns more than \$150,000
C. The average salary of the employees is between \$45,000 and \$145,000
D. More employees earn over \$100,000 than earn less than \$25,000

If f and g are functions such that $f(x) = (2x + 3)g(x)$, which of the following statements must be true?

A. The graph of $f(x)$ crosses the x-axis at $-\frac{2}{3}$

B. The graph of $f(x)$ crosses the x-axis at $-\frac{3}{2}$

C. The graph of $f(x)$ crosses the x-axis at $\frac{2}{3}$

D. The graph of $f(x)$ crosses the x-axis at $\frac{3}{2}$

If a is a constant, for what values of a does the line whose equation is $x + y = a(x - y)$ have a positive slope?

A. $a = 1$ B. $a \neq -1$ C. $-1 < a < 1$ D. $a < -1$ or $a > 1$

The graph whose equation is $(x - 4)^2 + (y - 2)^2 = 4$ is a circle. If m represents the number of times the circle intersects the y-axis and if n represents the number of times the circle intersects the x-axis, what is the value of $m + n$?

第六部分 考前5天复习方案

■ 答案与解析

1	C	【题干】：Joseph 加入一家俱乐部，收费为每月 \$79.99，并加上 6% 的税。同时他在参加该俱乐部的时候也要缴纳一项一次性费用为 \$95。由于没有合同所以他可以随时退出，不需要支付罚金。 【问题】：下面哪一个式子表达他成为该俱乐部会员 t 个月后的总费用？ 【解析】：根据题意可得 6% 的税是对应于每月的费用，因此 t 个月后费用为 \$79.99t 再加上税一共为：79.99t×(1+60%)，再加上单独的一次性费用总计为 79.99t×(1+60%)+95，整理后可得 1.06×79.99t+95，选 C。
2	16	【题干与问题】：已知 $x+y=5$ 和 $x^2+3xy+2y^2=40$，求 $2x+4y$ 的值。 【解析】：根据题意可得 $x^2+3xy+2y^2=(x+y)(x+2y)=40$，其中 $x+y=5$，得到 $x+2y=8$，因此 $2x+4y=16$。
3	B	【题干】：Paris 和 Genevieve 排队买玉米卷，周二的时候墨西哥快餐店会推出特价玉米卷：所有的玉米卷打 50% 的折扣。鱼肉玉米卷通常价格为 \$2.50/份，牛肉与鸡肉玉米卷通常价格为 \$1.50/份。他们正好需要 16 份玉米卷，但是消费又不能超过 \$15。 【问题】：他们最多能花多少钱买鱼肉玉米卷？ 【解析】：根据题意，设鱼肉玉米卷数量为 f，牛肉与鸡肉玉米卷数量为 b，可得关系式如下： $$f+b=16$$ $$2.5\times 50\%\times f+1.5\times 50\%\times b\leqslant 15$$ 式子 2 中可以取 15 来计算，得到 $f=6$，最后需要计算在鱼肉玉米卷上的花费：\$2.5×50%×f=\$2.5×50%×6=\$7.5，选 B。
4	C	【题干】：如果表达式 $\dfrac{9x^2}{3x+5}$ 可以写成 $\dfrac{25}{3x+5}+k$ 的形式。 【问题】：k 可以用什么式子来表示？ 【解析】：根据题意可得，$k=\dfrac{9x^2}{3x+5}-\dfrac{25}{3x+5}=\dfrac{9x^2-25}{3x+5}=\dfrac{(3x+5)(3x-5)}{3x+5}=3x-5$，选 C。
5	C	【题干】：一位农民在农场中随机挑选了 40 只鸭子称量体重。平均重量为 24 磅，误差幅度为 2.8 磅。该农民想重新调查一下以获得更小的误差幅度。 【问题】：下面哪一个样品可以导致更小的误差幅度？ A 项：从农场中随机挑选 25 只鸭子； B 项：从农场中随机挑选 25 只动物； C 项：从农场中随机挑选 100 只鸭子； D 项：从农场中随机挑选 100 只动物。 【解析】：根据题意可得，样本量将会影响最终的统计结果，样本量越小，误差越大。因此，要想获得更小的误差幅度，需要提高样本量，选 C。
6	37	【题干与问题】：根据给出的式子 $(2^{32})^{(2^{32})}=2^{(2^x)}$，求解 x 的值。 【解析】：根据题意，等式两边统一为以 2 为底，即 $32\times 2^{32}=2^x$，因为 $32=2^5$，所以 $2^x=2^5\times 2^{32}=2^{37}$，因此 $x=37$。

7	B	【题干】：哈特利家居用品店所有员工的年薪中位数值为 \$45,000，而员工收入的范围为 \$145,000。 【问题】：下面有关于中位数和范围的解释哪一项最合理？ 【解析】：本题需要解释中位数和范围，因此解题的关键是解释为什么有最大值（范围＝最大值－最小值，正是因为有了最大值，才导致了范围出现很大的值）。因此，选 B 项。值得注意的是，虽然 A 选项本身没有问题，但是不能解释中位数和范围。
8	B	【题干】：已知 f 和 g 是函数，并且满足 $f(x)=(2x+3)g(x)$。 【问题】：下面哪个陈述是正确的？ 【解析】：根据题意，问的是 $f(x)$ 与 x 轴的交点，本题虽然不知道 $g(x)$ 是多少，但是可以根据前面的式子来判断。根据定义找出 $f(x)$ 与 x 轴的交点（x，0）即是找出某一个 x 值，使得 $f(x)=0$。因此在本题中，可以得到当 $x=-\frac{3}{2}$ 时，$2x+3=0$，选 B。
9	D	【题干与问题】：a 是一个定值，下面哪一个值可以使得 $x+y=a(x-y)$ 的斜率为正值？ 【解析】：根据题意，将 $x+y=a(x-y)$ 转换可得 $y=\frac{a-1}{a+1}x$。要想获得斜率为正值，需要使得 $\frac{a-1}{a+1}>0$，解得 $a<-1$ 或 $a>1$，选 D。
10	1	【题干】：圆的方程为 $(x-4)^2+(y-2)^2=4$，m 代表圆与 y 轴的交点数量，n 代表圆与 x 轴的交点数量。 【问题】：$m+n$ 的值是多少？ 【解析】：从圆的方程 $(x-4)^2+(y-2)^2=4$ 可得圆的圆心坐标为（4，2），半径为2，如图 6-1 所示： 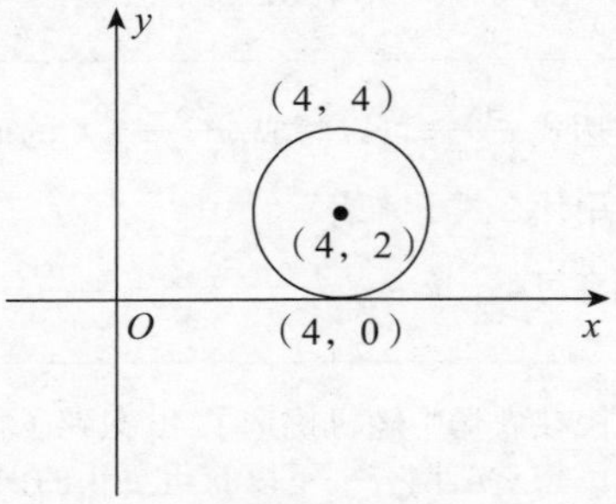 图 6-1 从图中可以看出该圆与 y 轴的交点数量为 0，与 x 轴的交点数量为 1，因此 $m+n=0+1=1$。

Day 3

Day 3 复习方案
复习内容
☐ 函数与数学模型
☐ 平面几何
☐ 三角函数
☐ 立体几何
考前适应性练习

The shaded region in the figure (Fig. 6-3) represents a rectangular frame with length 18 inches and width 15 inches. The frame encloses a rectangular picture that has the same area as the frame itself. If the length and width of the picture have the same ratio as the length and width of the frame, what is the length of the picture, in inches?

A. $9\sqrt{2}$

B. $\dfrac{3}{2}$

C. $\dfrac{9}{\sqrt{2}}$

D. $15\left(1-\dfrac{1}{\sqrt{2}}\right)$

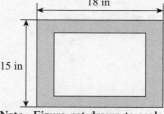

Note: Figure not drawn to scale

Fig. 6-3

If the figure (Fig. 6-4), if $AB /\!/ CE$, $CE = DE$, and $y = 45$, then $x =$
A. 45
B. 60
C. 67.5
D. 112.5

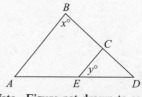

Note: Figure not drawn to scale

Fig. 6-4

A warehouse had a square floor with area 10,000 square meters. A rectangular addition was built along one entire side of the warehouse that increased the floor area by one-half as much as the original floor area. How many meters did the addition extend beyond the original building?

A. 10   B. 20   C. 50   D. 200

The figure (Fig. 6-5) shows the dimensions of a rectangular board that is to be cut into four identical pieces by making cuts at points A, B, and C, as indicated. If $x = 45$, what is the length of AB? (1 foot = 12 inches)

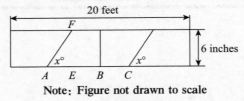

Note: Figure not drawn to scale

Fig. 6-5

A. 5 feet and 6 inches
B. 5 feet and 3 inches
C. 5 feet and $3\sqrt{2}$ inches
D. 5 feet

In the figure (Fig. 6-6), $\overline{BD}$ is tangent to circle A at point C and $\angle ABD = 25°$. The area of the shaded portion is exactly $\frac{1}{4}$ the area of circle A. What is $\angle CDE$ in degrees?

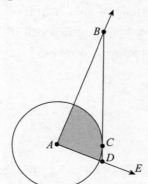

Note: Figure not drawn to scale

Fig. 6-6

The figure (Fig. 6-7) shows a rectangular solid with a length of 12 inches, a width of 4 inches, and a height of 6 inches. If points S and Q are midpoints of edges of the rectangular solid, what is the perimeter of the shaded region?

A. 14
B. 17.37
C. 34.74
D. 44.86

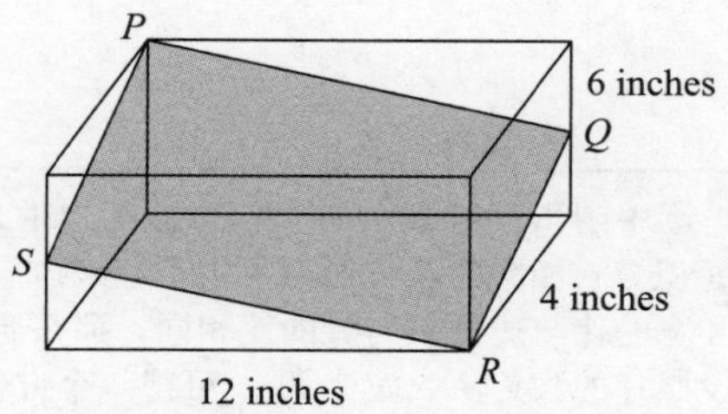

Fig. 6-7

Note: Figure not drawn to scale

In $\triangle ABC$, C is a right angle and $\tan A = 1$. What is the value of $\sin A + \cos A$?

A. $\frac{\sqrt{2}}{2}$ B. 1 C. $\sqrt{2}$ D. $2\sqrt{2}$

8

Two high school teachers took their classes on a field trip to a museum. One class spent $154 for admission for 20 students and 3 adults, and the other class spent $188 for admission for 24 students and 4 adults. Which of the following systems of equations could be solved to determine the price of a single student admission, s, and the price of a single adult admission, a, in dollars?

A. $a + s = 51$
   $44s + 7a = 342$

B. $20s + 3a = 154$
   $24s + 4a = 188$

C. $\frac{20}{s} + \frac{3}{a} = 154$
   $\frac{24}{s} + \frac{4}{a} = 188$

D. $20 + 24 = s$
   $3 + 4 = a$

The bird department of a pet store has 12 canaries, 30 finches, and 18 parrots. If the pet store purchased n more finches, then 80% of its birds would be finches. Which of the following equations must be true?

A. $\frac{1}{2} + n = \frac{4}{5}$

B. $\frac{30 + n}{60} = \frac{4}{5}$

C. $\frac{30 + n}{60 + n} = \frac{4}{5}$

D. $\frac{n}{60 + n} = \frac{4}{5}$

After its initial offering, the price of a stock increased by 20% in the first year, decreased by 25% in the second year, then increased by 10% in the third year. What was the net change in the stock price over the entire three-year period?

A. It increased by 5%

B. It increased by 1%

C. It decreased by 1%

D. It decreased by 5%

■ 答案与解析

1	A	【题干】：如图所示，图中阴影区域表示一个长方形的边框，长为 18 英寸，宽为 15 英寸。这样的边框把一个面积与它相同的长方形照片围了起来。 【问题】：若照片长和宽的比与边框长与宽的比相同，那么照片的长为多少英寸？ 【解析】：根据题意设照片的长为 L，宽为 W，可得一个方程组： $$2 \times L \times W = 18 \times 15$$ $$\frac{L}{W} = \frac{18}{15}$$ 解得 $L = 9\sqrt{2}$，选 A。
2	C	【题干】：如图所示，若 $AB \parallel CE$，$CE = DE$，且 $y = 45$。 【问题】：求 x 的值。 【解析】：由 $y = 45$，$CE = DE$，可得 $\angle ECD = \frac{180-45}{2} = 67.5°$，再根据 $AB \parallel CE$ 可得 $x° = \angle ECD = 67.5°$，选 C。
3	C	【题干】：一个仓库的地板是正方形，面积是 10,000 平方米。沿着该仓库的一条整边又额外建了一个长方形的房子，这样使得该仓库的面积比原来增加了一半。 【问题】：额外的房子伸出原来房子多少米？ 【解析】：根据题意，如图 6-2 所示： 设正方形地板的边长为 a，额外建的房子伸出原房子 b 米，则可以得到： $a^2 = 10,000$，解得 $a = 100$。由于该仓库的面积比原来增加了一半，因此 $a \times b = 5,000$，解得 $b = 50$，选 C。 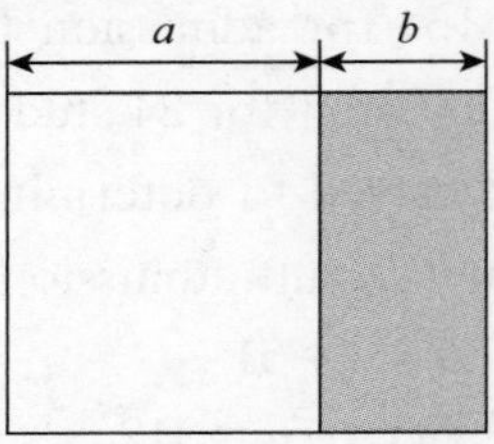 图 6-2
4	B	【题干】：如图 6-3 所示，一个长方形板的尺寸如图中所示，该长方形板将在图中的点 A，B，C 被切割成四块相同的板子。 【问题】：如果 $x = 45$，则 AB 的长度是多少？ 【解析】：本题暗含的一个条件需要注意，1 英尺 = 12 英寸。 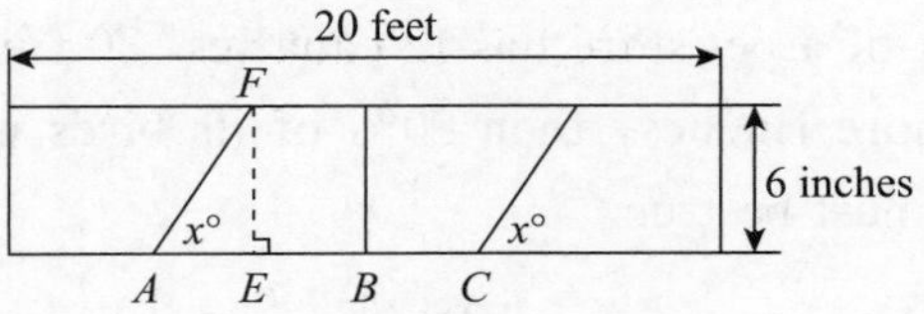 图 6-3 在图上做一条辅助线，其中 $EF \perp AB$，根据长方形的几何特征可知 $EF = AE = 6$ 英寸 = 0.5 英尺，$AC = 10$ 英尺，$BE = BC$，所以可得：$AC = AE + EB + BC = 2BE + AE = 10$，解得 $BE = BC = 4.75$，$AB = AE + BE = 4.75 + 0.5 = 5.25$ 英尺 = 5 英尺 3 英寸，选 B。
5	115	【题干】：如图所示，线段 $\overline{BD}$ 正切圆 A 于 C 点，$\angle ABD = 25°$，阴影部分的面积为四分之一圆。 【问题】：问 $\angle CDE$ 的角度是多少？ 【解析】：根据题意，由于阴影部分的面积为四分之一圆，所以可得 $\angle BAD = 90°$。因此 $\angle CDE = 180° - \angle ADB = 180° - (180° - \angle BAD - \angle ABD) = 180° - (180° - 90° - 25°) = 180° - 65° = 115°$。

第六部分　考前5天复习方案

6	C	**【题干】**：如图给出的是一个长方体，长为12英寸，宽为4英寸，高为6英寸。点 S 和 Q 是该长方体相应边的中点。 **【问题】**：阴影部分的周长是多少？ **【解析】**：根据题意可得，PS 和 QR 分别在两个全等的直角三角形中，其中两个直角边分别是3和4英寸，因此线段长度 $PS = QR = 5$ 英寸。PQ 和 SR 也分别在两个全等的直角三角形中，其中两个直角边分别是12和3英寸，因此线段长度 $PQ = SR = \sqrt{3^2 + 12^2} = \sqrt{153}$，因此 $PQRS$ 的周长为 $2 \times 5 + 2 \times \sqrt{153} \approx 34.74$，选 C。
7	C	**【题干】**：在三角形 ABC 中，C 是直角，$\tan A = 1$。 **【问题】**：$\sin A + \cos A$ 的值为多少？ **【解析】**：根据题意，由于 C 是直角，$\tan A = 1$，如图6-4所示：线段长度 $AC = BC$，可见该三角形是一个等腰直角三角形，即线段长度 $AC : BC : AB = 1 : 1 : \sqrt{2}$，因此 $\sin A + \cos A = \frac{1}{\sqrt{2}} + \frac{1}{\sqrt{2}} = \sqrt{2}$，选 C。 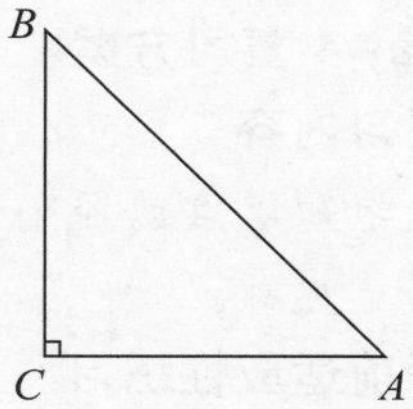 图 6-4
8	B	**【题干】**：两个高中老师带领学生去博物馆开展野外旅行。一个班级的入场费为 \$154，包括20名学生和3名成人，另外一个班级的入场费为 \$188，包括24名学生和4名成人。 **【问题】**：下面哪一个方程组表示一名学生的入场费 s 和一名成人的入场费 a？ **【解析】**：根据题意，一名学生的入场费为 s，一名成人的入场费为 a，一个班级的入场费为 \$154，包括20名学生和3名成人，因此可以写成 $20s + 3a = 154$。同理，另外一个班级的入场费为 \$188，包括24名学生和4名成人，因此可以写成 $24s + 4a = 188$，选 B。
9	C	**【题干】**：一家宠物店售鸟部有12只金丝雀、30只燕雀和18只鹦鹉。当该店购买了 n 只燕雀后，其80%的鸟为燕雀。 **【问题】**：下面哪一个式子是正确的？ **【解析】**：根据题意，原有30只燕雀，又买了 n 只燕雀后，一共是 $30 + n$ 只燕雀，因此一共有鸟的数量为 $12 + 30 + 18 + n = 60 + n$，可得 $(60 + n) \times 80\% = 30 + n$，化简为 $\frac{30 + n}{60 + n} = 80\% = \frac{4}{5}$，选 C。
10	C	**【题干】**：在首次发行之后，一股票的价格在第一年增加了20%，第二年下降了25%，第三年又增加了10%。 **【问题】**：在这三年中该股票的价格净变化为多少？ **【解析】**：根据题意，设该股票的原始价格为 x，变化情况为 $x \times (1 + 20\%) \times (1 - 25\%) \times (1 + 10\%) = 0.99x$。可见相对于原始价格，最终的价格为 $0.99x$，是下降了1%，选 C。

Day 4

Day 4 复习方案

复习内容

数学知识点已经全部复习完,今天的主要任务是通过练习几道典型例题来做考前的综合复习与热身。

考前适应性练习

If $f(n-1) = 13 + 4n$ for all values of n, what is the value of $f(3)$?

$PQRS$ is a rectangle (Fig. 6-8). The length of QR is 15, the length of QU is 10, and the length of SU is 13. What is the area of parallelogram $QUST$?

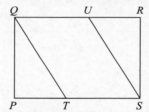

Note: Figure not drawn to scale

Fig. 6-8

Questions 3 through 4 refer to the following information.

An engineer performed strength-tests to measure the durability of a certain plastic. Fifty samples of the plastic were subjected to increasing pressures, and the results are presented below (Table 6-2):

Table 6-2

Pressure (psi)	# Broken	# Cracked
25	0	1
50	0	1
100	3	6
125	14	30
150	32	18

If 10,000 samples were subjected to 100 pounds per square inch (psi) of pressure, based on the results above, how many are expected to break?
A. 60   B. 300   C. 600   D. 1,200

At 125 pounds per square inch (psi), the break rate for sample sizes of fifty has a standard error (SE) of 0.06, and a critical value (CV) for a 95% confidence level of 1.96. Using the equation $ME = CV \times SE$, find the margin of error (ME) for the break rate of the entire population of samples at 125 psi.
A. 0.06   B. 0.12   C. 0.68   D. 0.95

A dessert recipe requires p tablespoons of sugar and q cups of flour. If Peter wants to make a larger batch using $p + 2$ tablespoons of sugar, how many cups of flour does he need to keep the ingredients in the original proportion?

A. $\dfrac{p}{(p+2)q}$ B. $\dfrac{p}{q}$

C. $\dfrac{p+2}{q}$ D. $\dfrac{(p+2)q}{p}$

A bungee jumper leaps off a cliff 122 meters from the ground. Her cord, when fully stretched, is 72 meters long, and it takes 6 seconds after the jump for it to extend fully. The distance between the jumper and the ground as a function of time can be modeled as a quadratic function. Which equation represents her distance from the ground as a function of time?
A. $f(t) = t^2 - 12t + 122$ B. $f(t) = 2t^2 - 24t + 122$
C. $f(t) = 2t^2 - 28t + 122$ D. $f(t) = t^2 - 12t + 122$

The equation of the parabola (Fig. 6-9) is $y = x^2$. If the y-coordinate of A and the y-coordinate of B are both 8, what is the length of $\overline{AB}$?
A. 16   B. 8
C. $2\sqrt{2}$ D. $4\sqrt{2}$

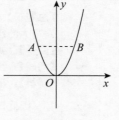

Fig. 6-9

Sixty cookies were equally distributed to x campers. Eight campers did not want cookies, so their share was redistributed to the other campers, who each received two more. What is the total number of campers?
A. 12   B. 20
C. 32   D. 40

Recent polls indicate that only 15% of those registered to vote in an upcoming election is in the age group 18 to 25. A voter registration drive wants to raise this figure to 20% before the day of the election, so it begins to target this demographic exclusively. If there are currently 51,000 registered voters in the district, and assuming all new registrations are in the target demographic. Approximately how many new voters does the drive need to register to meet its goal?

A. 2,600　　　　　B. 3,200　　　　　C. 6,800　　　　　D. 10,200

The gas mileage for Ken's car is 22 miles per gallon when his car is traveling at 60 miles per hour and is 25 miles per gallon when his car is traveling at 50 miles per hour. Ken will be driving from Boston to New York on a route that is 209 miles long. Gas costs $3.20 per gallon. How much more, in dollars rounded to the nearest cent, will Ken spend on gas if he drives the entire way at 60 miles per hour than if he drives the entire way at 50 miles per hour?

■答案与解析

1	29	【题干】：如果 $f(n-1) = 13 + 4n$。 【问题】：$f(3)$ 的值是多少？ 【解析】：根据题意，$f(3) = f(n-1)$，因此 $n = 4$，$13 + 4n = 29$。
2	120	【题干】：PQRS 是一个矩形，线段 QR 的长度为 15，线段 QU 的长度为 10，线段 SU 的长度为 13。 【问题】：平行四边形 QUST 的面积是多少？ 【解析】：根据题意可得，平行四边形的面积为 $A = bh$，从图中可知平行四边形的高为线段 RS，在直角三角形 RSU 中，符合 5—12—13 的勾股定理关系，因此可得线段 RS 的长度 = 12。因此平行四边形的面积为 $A = bh = 10 \times 12 = 120$。
3	C	【总题干】：一位工程师通过进行强度试验来衡量某些塑料的耐久性，50 个塑料样品受到不断增大的压力作用，结果如表中所示。 【题干】：如果 10,000 个样本受到 100 磅/平方英寸（psi）的压力作用。 【问题】：预测有多少个样品会损坏？ 【解析】：根据题意可得：当给的压力为 100 磅/平方英寸时，50 个塑料样品中会有 3 个损坏。因此设 10,000 个样品中损坏的个数为 x，可得：$\frac{x}{10,000} = \frac{3}{50}$，解得 $x = 600$，选 C。
4	B	【题干】：当给的压力为 125 磅/平方英寸时，50 个样品损坏率的标准误差为 0.06，95% 置信区间的临界值为 1.96。使用公式 $ME = CV \times SE$ 来计算误差界限（ME）。 【问题】：所有样品在压力为 125 磅/平方英寸时的误差界限为多少？ 【解析】：根据题意，误差界限 = 标准误差 × 临界值 = $0.06 \times 1.96 = 0.12$，选 B。

第六部分 考前5天复习方案

5	D	【题干】：一份甜点食谱需要 p 汤匙的糖和 q 杯面粉。如果 Peter 需要制作一份更大的甜点，其中需要 $p+2$ 汤匙的糖。 【问题】：他需要加多少杯面粉才能保持原有的配比？ 【解析】：根据题意，设需要加入 x 杯面粉，要保持原有的配比，需要符合以下等式：$\frac{p}{q} = \frac{p+2}{x}$，解得 $x = \frac{(p+2)q}{p}$，选 D。
6	B	【题干】：蹦极跳爱好者从离地面 122 米高的悬崖跳下，如果其身上捆绑的绳索完全伸长为 72 米。从他开始跳到绳索完全伸长需要 6 秒。这位蹦极跳爱好者与地面之间距离与时间的关系可以用一个二次函数来表示。 【问题】：下面哪一个式子表示的是这个关系？ 【解析】：根据题意，悬崖高 122 米，绳索最长为 72 米。当绳索完全伸长的时候，他离地面的高度是最短的，此时所用时间为 6 秒。由于需要满足二次函数，可以写成顶点式：$f(t) = a(t-6)^2 + 122 - 72$，对照选项可以发现，符合这个顶点式的只有 B 项。
7	D	【题干】：如果一个抛物线的方程为 $y = x^2$，A 和 B 的纵坐标都为 8。 【问题】：线段 AB 的长度是多少？ 【解析】：根据题意，要想求线段 AB 的长度，即求解 A 和 B 横坐标，可得到方程 $8 = x^2$，解得 $x = 2\sqrt{2}$，因此线段 AB 的长度 $= 2\sqrt{2} \times 2 = 4\sqrt{2}$，选 D。
8	B	【题干】：60 份曲奇平均分配给 x 位露营者，有 8 位露营者不需要曲奇，因此他们的曲奇可以再次分配给其他露营者，使得这些露营者每人多分了 2 个。 【问题】：一共有多少位露营者？ 【解析】：根据题意：60 份曲奇平均分配给 x 位露营者，每位可分得 $\frac{60}{x}$，由于 8 位不要，因此多出来 $8 \times \frac{60}{x}$ 份，再分给剩下的 $x-8$ 位，每人多分得 2 份即 $$8 \times \frac{60}{x} = 2 \times (x-8)$$ 解得 $x = 20$ 或 -12，取正值，因此一共有 20 位露营者，选 B。
9	B	【题干】：最近的民调显示在即将到来的选举中，年龄 18 到 25 岁的选民只占登记投票人数的 15%。一个选民登记运动组织希望在选举日之前将这一比例提高到 20%。因此该组织开始针对这一部分人群做工作。如果该地区现有登记选民为 51,000，假设新增的选民都是在这个年龄段的。 【问题】：如果要达到希望的目标，这个选民登记运动组织需要动员多少新的选民？ 【解析】：根据题意，原有 18 至 25 岁选民有 $51,000 \times 15\% = 7,650$，设新增的选民数为 x，题中假设新增的选民都是在这个年龄段，因此可得 $\frac{7,650+x}{51,000+x} = 0.2$，解得 $x = 3,188 \approx 3,200$，选 B。
10	3.65	【题干】：当以 60 英里/小时的速度行驶时，Ken 轿车的油耗为 22 英里/加仑。而以 50 英里/小时的速度行驶时，油耗为 25 英里/加仑。Ken 从波士顿开车去纽约，路程为 209 英里，油价为每加仑 \$3.20。 【问题】：Ken 以 60 英里/小时的速度行驶比以 50 英里/小时的速度行驶需要多花费多少油钱？（结果保留到百分位） 【解析】：根据题意，全程为 209 英里，如果以 60 英里/小时的速度行驶，需要 $\frac{209}{22} = 9.5$ 加仑的汽油，如果以 50 英里/小时的速度行驶，需要 $\frac{209}{25} = 8.36$ 加仑的汽油，因此多出了 $9.5 - 8.36 = 1.14$ 加仑的汽油，因此多花了 $\$3.20 \times 1.14 = \$3.648 \approx \$3.65$。

Day 5

Day 5 复习方案

1. 将本书中所有的错题再次复习一遍。
2. 再次熟悉一下数学词汇。
3. 将考试所需要的文具检查一下，尤其是做题所用的 2B 铅笔与计算器、橡皮，再次确认没有问题。
4. 检查考试证件、适当的食品和饮料。注意不要违反考试规则，不做和考试无关的事。
5. 多关注自己的邮箱，及时关注考试资讯。
6. 早点休息，养足精神。

错题记录

序号	页码	题号	主要问题与错误分析

附录1 知识点清单

1. 数与数的运算

知识点

1. 数列（sequence）：一组有顺序的数字组，常见的数列有：

等差数列（arithmetic sequence）：前后两项的差是定值，其中第 n 项 a_n 的通式可以写成 $a_n = a_1 + (n-1) \times d$，其中 a_1 为首项，d 为公差。

等比数列（geometric sequence）：相邻两项之间成一个固定的比例，其中第 n 项 a_n 的通式可以写成 $a_n = a_1 \times r^{n-1}$，其中 a_1 为首项，r 为公比。

2. 集合：集合是某些内容的聚集，这些内容称为该集合的元素。常见的集合有并集和交集。

并集：两个集合中所包含的所有元素，可以表示为 $A \cup B$。

交集：两个集合中的共有元素，可以表示为 $A \cap B$。

3. 复数（complex number）：我们把形如 $a + bi$（a，b 均为实数）的数称为复数，其中 a 称为实部，b 称为虚部，i 称为虚数单位（人为设定 $i^2 = -1$，即 $i = \sqrt{-1}$）。当虚部等于零时，这个复数可以视为实数；当 z 的虚部不等于零、实部等于零时，常称 z 为纯虚数。两个实部相等，虚部互为相反数的复数互为共轭复数，如 $a + bi$ 和 $a - bi$。复数的运算法则包括：加减法、乘法以及分母有理化。例如，设 Z_1 和 Z_2 为两个复数，即 $Z_1 = a + bi$，$Z_2 = c + di$，则有：

$Z_1 + Z_2 = (a + c) + (b + d)i$

$Z_1 - Z_2 = (a - c) + (b - d)i$

$Z_1 \times Z_2 = (a + bi)(c + di) = ac + adi + bci + bdi^2 = (ac - bd) + (ad + bc)i$

分母有理化是指当出现 $\dfrac{a + bi}{c + di}$ 的时候，通过分子、分母同时乘以分母的共轭复数而将分母中的 i 化简掉，即 $\dfrac{a + bi}{c + di} = \dfrac{(a + bi) \times (c - di)}{(c + di) \times (c - di)} = \dfrac{(ac + bd) + (bc - ad)i}{c^2 + d^2} = \dfrac{ac + bd}{c^2 + d^2} + \dfrac{bc - ad}{c^2 + d^2} \times i$。

过关测试

1.   50，44.5，39，33.5，…

The tenth number of the sequence is:
A. −4　　　　B. 0.5　　　　C. 1　　　　D. 1.5

2. What is the next term in the sequence: 2, 8, 32, 128, ⋯?

3. The first term in a geometric sequence is 2, and the common ratio is 3. The first term in an arithmetic sequence is 3, and the common difference is 3. Let set X be the set containing the first six terms of the geometric sequence and set Y be the set containing the first six terms of the arithmetic sequence. What is the sum of the elements in $X \cap Y$?

4. Which of the following choices could be equal to set Z if $X = \{2, 5, 6, 7, 9\}$ and $Y = \{2, 5, 7\}$?
$X \cup Y \cup Z = \{1, 2, 3, 4, 5, 6, 7, 8, 9\}$, $X \cap Z = \{2, 6\}$, $Y \cap Z = \{2\}$
A. $Z = \{1, 4, 8\}$　　　　　　　B. $Z = \{1, 3, 4, 8\}$
C. $Z = \{1, 2, 3, 4, 6, 8\}$　　　D. $Z = \{1, 2, 3, 5, 6, 8\}$

5. If $i^2 = -1$, and $(4+2i)(6-ki) = 30$, what is the value of k?
A. 3　　　　B. 4　　　　C. 6　　　　D. 8

6. If A is the solution set of the equation $x^2 - 4 = 0$ and B is the solution set of the equation $x^2 - 3x + 2 = 0$, how many elements are in the union of the two sets?

■答案与解析

1	B	【题干与问题】：该数列的第10项为多少？ 【解析】：该数列是以5.5递减，根据公式可得 $a_{10} = a_1 + (10-1) \times 5.5 = 0.5$，选B。
2	512	【题干与问题】：该数列的下一项为多少？ 【解析】：该数列为等比数列，公比为4，因此下一项应该为 $128 \times 4 = 512$。
3	24	【题干】：一个等比数列的首项为2，公比为3。一个等差数列的首项为3，公差为3。设集合 X 为该等比数列的前6项，集合 Y 为该等差数列的前6项。 【问题】：$X \cap Y$ 中元素之和为多少？ 【解析】：由题意可得： set $X = \{2, 6, 18, 54, 162, 486\}$, set $Y = \{3, 6, 9, 12, 15, 18\}$，因此 $X \cap Y = \{6, 18\}$，该集合中所有元素求和，得到 $6 + 18 = 24$。
4	C	【题干与问题】：下列哪一个选项等于集合 Z？ 【解析】：由于集合 X 和 Y 中均不含4，根据 $X \cup Y \cup Z = \{1, 2, 3, 4, 5, 6, 7, 8, 9\}$，说明 Z 中含有4。再根据 $X \cap Z = \{2, 6\}$ 和 $Y \cap Z = \{2\}$，可得 Z 中有6。符合这一条件的只有C项。
5	A	【题干】：假设 $i^2 = -1$ 并且 $(4+2i)(6-ki) = 30$。 【问题】：k 的值为多少？ 【解析】：根据题意，将 $(4+2i)(6-ki)$ 展开得到 $24 + 2k + (12-4k)i = 30$。当虚部为零时，这个复数可以视为实数。因此 $12 - 4k = 0$，或者 $24 + 2k = 30$，解得 $k = 3$，选A。
6	3	【题干】：A 是方程 $x^2 - 4 = 0$ 解的集合，B 是方程 $x^2 - 3x + 2 = 0$ 解的集合。 【问题】：这两个集合共有多少个元素？ 【解析】：根据题意可以得到 $A = \{-2, 2\}$，$B = \{1, 2\}$，因此两个集合有 −2，1 和 2 共计 3 个元素。

2. 因子与倍数

知识点

1. 质数/素数（prime）：一个大于1的自然数，除了1和它本身没有别的约数，那么这个数就为质数，否则称为合数。最小的质数为2，最小的合数为4，1既不是素数，也不是合数。

2. 整数（integer）：整数包括正整数（positive integers）、负整数（negative integers）和零。

3. 偶数（even integers）：指能被2整除的整数，可以表示为形如$2n$的数（n为整数），而奇数（odd integers）可以表示为$2n+1$的数（n为整数）。注意：0是偶数。

4. 整除（divisibility）：一个整数能够被另一整数整除，这个整数就是另一整数的倍数（multiples）。请注意表达：如，35能够被5或7整除，或者说5或7能够整除35。

5. 最大公约数（greatest common factor，GCF）：指两个或多个整数共有约数中最大的一个。如：34和52的最大公约数为2。

6. 最小公倍数（least common multiple，LCM）：两个或多个整数公有的倍数叫做它们的公倍数，其中两个或多个整数的公倍数里最小的那一个叫做它们的最小公倍数。

7. 合数（composite）：能被非1及其本身的某一个或多个素数整除的自然数，即含有两个以上的因子。

过关测试

1. Let P be a prime number greater than 4. How many distinct prime factors does $9 \times P^2$ have?
A. 2　　　　　　B. 3　　　　　　C. 4　　　　　　D. 5

2. How many positive integers less than 20 have an odd number of distinct factors?
A. 12　　　　　　B. 10　　　　　　C. 8　　　　　　D. 4

3. In the repeating decimal 0.714 285 714 285…, what is the 50th digit to the right of the decimal point?

4. How many distinct composite numbers can be formed by adding 2 of the first 5 prime numbers?

5. k divided by 4 has a remainder of 3. What is the remainder when $k+3$ is divided by 4?

■答案与解析

1	A	【题干】：假设 P 是一个大于4的素数。 【问题】：$9 \times P^2$ 有多少个不相同的素因子？ 【解析】：根据题意可得：$9 \times P^2$ 可以写成 $3 \times 3 \times P \times P$，可见一共有四个素因子，其中有两个相同，因此不相同的因子有2个，分别为3和P，选A。

2	D	【题干与问题】：小于 20 的正整数中有多少个数是有奇数个不相同因子的？ 【解析】：根据题意：从小于 20 的正整数中挑选需要注意： 1）因子要不相同。2）因子的数量为奇数个。 有如下几个： 1：1； 4：1，2，4； 9：1，3，9； 16：1，2，4，8，16。 一共有 4 个，选 D。
3	1	【题干与问题】：在一个重复的小数 0.714 285 714 285…中，小数点后的第 50 位数是多少？ 【解析】：根据题意可得：0.714 285 714 285… 发现 "714285" 正好是六位形成一组在不断重复。因此第 50 位正好是八次重复之后再往右移动两位，为 "1"。
4	7	【题干与问题】：前五个素数两两相加可以形成多少个不相同的复合数？ 【解析】：根据题意可得，前五个素数为 2，3，5，7，11，两两相加为： $2+3=5$，$2+5=7$，$2+7=9(\checkmark)$，$2+11=13$ $3+5=8(\checkmark)$，$3+7=10(\checkmark)$，$3+11=14(\checkmark)$ $5+7=12(\checkmark)$，$5+11=16(\checkmark)$ $7+11=18(\checkmark)$ 一共有 7 个。
5	2	【题干与问题】：k 除以 4 余数为 3，问 $k+3$ 除以 4 的余数为多少？ 【解析】：这样的题可以考虑假设代入法，因此假设 $k=7$，满足 k 除以 4 余数为 3，则 $k+3=10$，10 除以 4 商为 2，余数为 2。

3. 比例

知识点

1. 比率（ratio）：表示两个量的数学关系，可以表示成两种形式，如：3 to 4，或 3：4。

2. 如果 y 随 x 成比例变化，称为 y is proportional to x，如果 $y = kx$，称为正比例（directly proportional）。如果 $y = \dfrac{k}{x}$，称为反比例（inversely proportional），其中 k 均为常数，且 $k \neq 0$。

过关测试

1. 3 gallons of paint are needed for a wall that is 100 ft^2. How many gallons of paint must be purchased at the store for 480 ft^2 of wall space?

A. 14 gallons   B. 15 gallons   C. 16 gallons   D. 17 gallons

2. A company produces baseballs at 3 different plants：Plant A，Plant B，and Plant C. Doubling Plant A's production is equal to $\dfrac{1}{3}$ of the company's total production. Doubling Plant B's production is equal to the company's total production. Tripling Plant C's production is equal to the company's total production. What is the ratio of Plant A's production to Plant

B's production to Plant C's production?

A. 1∶3∶2   B. 3∶1∶2   C. 2∶1∶3   D. 1∶2∶3

3. If Greg lost 20 pounds, then the ratio of Ted's weight to Greg's weight would be $\frac{4}{3}$. If Ted weighs 180 pounds, what was Greg's initial weight?

A. 115 pounds   B. 125 pounds   C. 135 pounds   D. 155 pounds

4. Company A, Company B, and Company C are three Internet providers in a certain area of the country. The ratio of subscribers of A to B to C is 2∶5∶6. If there are a total of 65,000 subscriptions, how many of the 65,000 use Company A and Company C?

A. 55,000   B. 40,000   C. 35,000   D. 30,000

5. Ethanol is an alcohol commonly added to gasoline to reduce the use of fossil fuels. A commonly used ratio of ethanol to gasoline is 1∶4. Another less common and more experimental additive is methanol, with a typical ratio of methanol to gasoline being 1∶9. A fuel producer wants to see what happens to cost and fuel efficiency when a combination of ethanol and methanol are used. In order to keep the gasoline used the same, what ratio of ethanol to methanol should the company use?

A. 1∶1   B. 4∶9   C. 9∶4   D. 36∶9

6. If the ratio of boys to girls at a school picnic is 5∶3, which of the following CANNOT be the number of children at the picnic?

A. 24   B. 40   C. 96   D. 150

■ 答案与解析

1	B	【题干】：3 加仑的油漆可以用于 100 平方英尺的墙。 【问题】：480 平方英尺的墙需要多少加仑的油漆？ 【解析】：根据题意，设需要 x 加仑的油漆，可得：$\frac{x}{480} = \frac{3}{100}$ 解得 $x = 14.4$，注意本题中问的是需要多少加仑的油漆，所以需要进位到 15，选 B。
2	A	【题干】：一个公司拥有三家生产棒球的工厂 A，B 和 C。A 工厂产量的两倍为公司总量的 $\frac{1}{3}$，B 工厂产量的两倍为公司总量，C 工厂产量的三倍为公司总量。 【问题】：A 工厂、B 工厂和 C 工厂的产量之比为多少？ 【解析】：根据题意，设 A 工厂的产量为 A，B 工厂的产量为 B，C 工厂的产量为 C，公司的总产量为 T，则满足以下关系式： $$2A = \frac{1}{3}T$$ $$2B = T$$ $$3C = T$$ 因此可以得到 $A = \frac{1}{6}T$，$B = \frac{1}{2}T$，$C = \frac{1}{3}T$，A 工厂、B 工厂和 C 工厂的产量之比即 $\frac{1}{6}T : \frac{1}{2}T : \frac{1}{3}T = 1:3:2$，选 A。

3	D	【题干】：如果 Greg 减肥 20 磅，那么 Ted 和 Greg 的体重之比为 $\frac{4}{3}$。Ted 的体重为 180 磅。 【问题】：Greg 原体重为多少？ 【解析】：根据题意：设 Greg 原体重为 G 磅，Ted 的体重为 T 磅，则满足以下关系式： $$\frac{4}{3}=\frac{T}{G-20}$$ 解得 $G=155$，选 D。
4	B	【题干】：A，B 和 C 公司是某一地区网络提供商，三个公司的客户量之比为 2：5：6。总客户量为 65,000。 【问题】：A 和 C 公司的客户量为多少？ 【解析】：根据题意，设 A，B 和 C 公司的客户量分别为 $2x$，$5x$ 和 $6x$，可得以下关系式：$2x+5x+6x=65,000$，解得 $x=5,000$，A 和 C 公司的客户量为 $2x$ 加 $6x$，即为40,000，选 B。
5	C	【题干】：为了减少化石燃料的使用，通常将乙醇添加到汽油中。一般而言，乙醇与汽油的比例是 1：4。甲醇是另外一种添加剂，甲醇与汽油的比例是 1：9。一个燃料生产商希望了解一下将乙醇和甲醇添加在一起后燃料的成本与效率。 【问题】：如果要保持汽油与添加剂之间的比例是相同的，该公司需要采用的两种添加剂的比例是多少？ 【解析】：根据题意可以发现，乙醇（e）与汽油（g）的比例 $e：g=1：4$，甲醇（m）与汽油（g）的比例 $m：g=1：9$，如果要保证汽油的用量一致，即需要找到"1：4"和"1：9"中"4"和"9"的公倍数，可以得到 $e：g=1：4=9：36$，$m：g=1：9=4：36$。因此 $e：m：g=9：4：36$，乙醇和甲醇的比例是 9：4，选 C。
6	D	【题干】：在学校的一次野餐活动中男孩和女孩的比例为 5：3。 【问题】：下面哪一个数字不可能是此次野餐活动中孩子的数量？ 【解析】：根据题意可以发现，设 $5x$ 和 $3x$ 分别为男孩和女孩的数量，总数则就为 $8x$。由于本题没有给出具体的数据，无法直接计算出答案，因此需要通过数据的"形式"来判断。可见不论总数为多少，都应该总数是 8 的倍数，从四个选项中可以看出 D 项 150 不是 8 的倍数，因此总数不可能为 150，选 D。

4. 百分比

知识点

1. 百分数（percent,%）：分母是 100 的分数，是在比率式中第二个量为 100 时的比率。

2. 增加和下降的百分比：

增加百分比（percent increase）$=\dfrac{\text{新的数值}-\text{起始数值}}{\text{起始数值}}\times 100\%$

下降百分比（percent decrease）$=\dfrac{\text{起始数值}-\text{新的数值}}{\text{起始数值}}\times 100\%$

3. 常见百分比的几种表达方式（表附 1）：

表附1

表述方式	数学表达式
$k\%$ of A is B	$A \times k\% = B$
What percent of A is B	$\dfrac{B}{A} \times 100\%$
A increased by $k\%$ is B	$A \times (1 + k\%) = B$

过关测试

1. A camping tent went on sale from $80 to $60. A camping chair that originally cost $20 was dis-counted by the same percent. What is the new price of the camping chair?
A. 25　　　　　　　　　　　B. 20
C. 15　　　　　　　　　　　D. 10

2. Dan needs to gain 8% of his current body weight to wrestle in weight Class A. Dan needs to lose the same percent of his current weight to wrestle in weight Class B. What percent of Dan's Class A weight, to the nearest tenth, does he need to lose to get to his Class B weight?

3. During last year's baseball season, a certain player has 625 at bats and gets a hit 32% of the time. This season the player increased the at bats by 12%, and got a hit 34% of the time. What is the percent increase in the number of hits?

4. The table (Table A1) shows the results of a survey about high school students' support for reporting class rank to colleges. How many non-seniors do not support reporting class rank to colleges?

Table A1

	Number of students	Percent Who Support Reporting Class Rank
All students	800	60%
Seniors	225	80%

5. The table (Table A2) shows the results of a sociological study identifying the number of males and females with and without college degrees who were unemployed or employed at the time of the study. If one person from the study is chosen at random, what is the probability that the person is an employed person with a college degree?

A. $\dfrac{73}{160}$　　　　　　　　　　　B. $\dfrac{10}{17}$

C. $\dfrac{17}{20}$　　　　　　　　　　　D. $\dfrac{73}{80}$

Table A2

	Unemployed	Employed	Totals
Female Degree	12	188	200
Female No Degree	44	156	200
Male Degree	23	177	200
Male No Degree	41	159	200
Totals	120	680	800

6. Brian gave 20% of his baseball cards to Scott and 15% to Adam. If he still had 520 cards, how many did he have originally?

■ 答案与解析

1	C	【题干】：一个野营帐篷的价格从 $80 下降到 $60。野营椅子的原价为 $20，也按照同样的幅度打折降价。 【问题】：野营椅子现在的新价格是多少？ 【解析】：根据题意，野营帐篷的价格下降的百分比为： $$\frac{\$80-\$60}{\$80}\times100\%=25\%$$ 因此，野营椅子也是按照这样的折扣降价，可得新的价格为 $20×(1-25%)= $15，选 C。
2	14.8	【题干】：在摔跤比赛中，Dan 现在的体重增加 8% 后他可以划归到 A 级别，而现在的体重下降同样的百分比后他可以划归到 B 级别。 【问题】：Dan 的体重从 A 级别到 B 级别下降的百分比是多少？（结果保留到十分位） 【解析】：根据题意，本题中没有给出体重具体的数值，只是给出了百分比，因此可以假设 Dan 现在的体重为 100，这样便于后续的计算： Dan 如果划归到 A 级别的体重为 100×(1+8%)=108 Dan 如果划归到 B 级别的体重为 100×(1-8%)=92 可见 Dan 的体重从 A 级别到 B 级别下降的百分比是 $\frac{108-92}{108}\times100\%\approx14.8148\%\approx14.8\%$。
3	19	【题干】：在去年的棒球赛季中，某个球员用球棒击球 625 次，其中安打为 32%。这个赛季这个球员击球率增加了 12%，其中安打为 34%。 【问题】：这个球员安打提高的百分比是多少？ 【解析】：根据题意可得：本赛季击球率为 625×(1+12%)=700，其中安打为 700×34%=238。而在上个赛季，安打为 625×32%=200，因此安打提高的百分比是 $\frac{238-200}{200}\times100\%=19\%$。 (安打：get a hit，是棒球及垒球运动中的一个名词，指打击手把投手投出来的球，击出到界内，使打者本身能至少安全上到一垒的情形。)
4	275	【题干】：下面的表格是对高中学生的一项调查，内容有关于是否支持向大学提供班级排名。 【问题】：有多少非高年级学生不支持向大学提供班级排名？ 【解析】：根据题意可得，有 800 名学生，不支持的人数为 800×(1-60%)=320。高年级学生为 225 人，其中不支持的人数为 225×(1-80%)=45，因此非高年级学生不支持的人数为 320-45=275。

5	A	**【题干】**：表中显示了一项社会学研究，结果为男性和女性的数量以及是否有大学学位的人失业或就业情况，如果研究对象是随机选取的。 **【问题】**：抽到一个人，这个人是有工作且拥有大学学位的概率是多少？ **【解析】**：根据题意，从表中看出总人数为 800 人，其中女性有大学学位和有工作的人数为 188，男性有大学学位和有工作的人数为 177，因此满足问题条件的人数为 188 + 177 = 365 人。抽到这类人的概率为 $\frac{365}{800} = \frac{73}{160}$，选 A。
6	800	**【题干】**：Brian 把他持有的棒球卡数量的 20% 给了 Scott，15% 给了 Adam，如果他现在仍然有 520 张卡。 **【问题】**：他原来有几张卡？ **【解析】**：设他原来有 x 张卡，根据题意可得：Brian 一共给出了 20% + 15% = 35% 的卡，因此剩余卡占原来总数的 1 - 35% = 65%。可得 $x \times 65\% = 520$，解得 $x = 800$。

5. 统计与概率

知识点

1. 平均数（mean）：由一组数据的总和除以这组数据的个数后得到的数是平均数，平均数只有 1 个。

统计意义：反映了一组数据的平均大小，常用来代表数据的总体"平均水平"。

2. 众数（mode）：在一组数据中出现次数最多的数，不必计算就可求出。值得注意的是在一组数据中，可能不止一个众数，也可能没有众数。

统计意义：反映了出现次数最多的数据，用来代表一组数据的"多数水平"。

3. 中位数（media）：首先将一组数据按大小顺序排列，处在最中间位置的一个数（奇数个原始数据）或最中间位置两个数的算术平均值（偶数个原始数据）。

统计意义：中位数像一条分界线，将数据分成前半部分和后半部分，中位数代表一组数据的"中等水平"。

4. 加权平均数（weight mean/average）：加权平均值即将各数值乘以相应的权数，然后加总求和得到总体值，再除以总的单位数。公式为：

$$(X_1 \times f_1 + X_2 \times f_2 + X_3 \times f_3 + \cdots + X_n \times f_n)/(f_1 + f_2 + f_3 + \cdots + f_n)$$

其中 $X_1 \sim X_n$ 代表的是具体的数字，$f_1 \sim f_n$ 代表的是对应的数字所占的比重（叫做权重）。

5. 概率（probability）：一件事情发生的可能性称为该事件的概率，计算公式如下：

$$P = \frac{该事件发生的次数}{所有事件发生的次数}$$

6. 排列（permutation）：如果从一组数中的 n 个元素中一个接一个地挑选 m 个元素，则选择的可能数在每次选择之后会减少 1，这就是排列问题，记作 $P_n^m = \frac{n!}{(n-m)!}$。

7. 组合（combinations）：如果从同一组数中的 n 个元素中挑选 m 个元素，挑选过程没有顺序要求，这就是组合问题，记作 $C_n^m = \frac{n!}{m!(n-m)!}$。

8. 抽取的有放回与无放回两种情况：例如现有一批产品共 20 件，其中正品 12 件，次品为 8 件，如果一次性抽取 3 件，求抽取的 3 件均为正品的概率：

正品一共 12 件，选到 3 件的组合有 C_{12}^3，总体有 C_{20}^3 种抽样可能，所以发生的概率为 $\dfrac{C_{12}^3}{C_{20}^3}$，如果抽取一件还要放回，则每次抽取到正品的概率是 $\dfrac{12}{20}$，三次的概率为 $\dfrac{12}{20} \times \dfrac{12}{20} \times \dfrac{12}{20}$。

9. 维恩图（Venn diagram）：采用固定位置的交叉环形式，用封闭曲线来表示集合及其关系，如图附 1 所示：

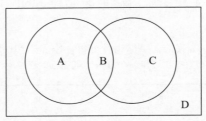

图附 1

A 区域代表仅满足 A 条件，C 区域代表仅满足 C 条件，B 区域代表既满足 A 条件、也满足 C 条件。D 区域代表既不满足 A 条件，也不满足 C 条件。

过关测试

Questions 1 and 2 refer to the following information.

The tables (Table A3) show the points scored by two different basketball players, Jay and Ed. Ed's average (arithmetic mean) for 4 games is 2 less than Jay's average for 5 games.

Table A3

Jay' points	Ed's points
20	15
16	21
25	17
15	
14	

1. How many points did Ed score during the fourth game?
A. 12　　　　B. 11　　　　C. 10　　　　D. 9

2. Say that Ed scored 16 points in the fourth game. What should be added to Jay's median score to equal the median of Ed's scores?

3. What is the median of the modes in the data set $\{-5, 4, 3, 7, 2, 1, 3, 4, 5, -1, 7, 8, -4, 2, 6\}$?
A. 2　　　　B. 2.5　　　　C. 3　　　　D. 3.5

4. s and t are positive integers whose average (arithmetic mean) is 9. If $s < t$, what is the median of all possible values of s?

5. The modes of a set of 9 numbers are x, y, and z, and the average (arithmetic mean) of the 9 numbers is 20. Three of the 9 numbers are $2x+5$, $2y$, and $2z-3$. What is the value of $4(x+y+z)$?

A. 178   B. 179   C. 180   D. 181

6. How many positive 4-digit numbers are there with an even digit in the hundreds position and an odd digit in the tens position?

A. 10,000   B. 5,040   C. 2,500   D. 2,250

7. How many 5-digit numerals have 9 as the first digit, 3 or 6 as the third digit, and no digit repeated?

8. Three different integers are randomly selected from a group of five unique integers consisting of 1 through 5. What is the probability that these number are 1, 2 and 3?

■答案与解析

1	B	【题干】：表中是 Jay 和 Ed 这两位篮球运动员的投篮成绩。Ed 在四场比赛的平均得分比 Jay 五场比赛平均得分要少 2 分。 【问题】：Ed 在第四场比赛的得分是多少？ 【解析】：根据题意，设 Ed 在第四场比赛的得分是 x，满足以下式子 $\frac{15+21+17+x}{4} = \frac{20+16+25+15+14}{5} - 2$，解得 $x=11$，选 B。
2	0.5	【题干】：假设 Ed 的第四场得分为 16。 【问题】：Jay 得分的中位数需要加上多少才能与 Ed 得分的中位数相等？ 【解析】：根据题意，Ed 的第四场得分为 16，因此四场得分从小到大排列为 15，16，17，21，因此其成绩的中位数为 16.5，而 Jay 得分的中位数为 16。可见还需要加上 16.5 - 16 = 0.5 分。
3	D	【题干与问题】：该集合众数的中位数是多少？ 【解析】：根据题意可得：该集合的众数为 2，3，4，7，因此众数的中位数为 3.5，选 D。
4	4.5	【题干】：如果 s 和 t 都是正整数，它们的平均值为 9，且 s 小于 t。 【问题】：所有 s 可能值的中位数是多少？ 【解析】：根据题意，s 和 t 都是正整数，平均值为 9，因此 $s+t=18$，本题转化为要找两个正整数之和为 18 的几种情况。有以下几种情况（从小往大排列）： (1, 17)；(2, 16)；(3, 15)；(4, 14)；(5, 13)；(6, 12)；(7, 11)；(8, 10)。注意由于题目要求 s 小于 t，因此从 (9, 9) 开始及其之后的组合就不满足条件了。满足条件的所有 s 值可以形成一个集合：$\{1, 2, 3, 4, 5, 6, 7, 8\}$，可得中位数为 4.5。
5	A	【题干】：一个含九个数字的集合其众数为 x，y 和 z，九个数字的平均值为 20。其中三个数字分别为 $2x+5$，$2y$ 和 $2z-3$。 【问题】：$4(x+y+z)$ 的值为多少？ 【解析】：根据题意，众数是该集合中同时出现频次最多且相等的那个或者那一些数字。原题给出的条件是该集合的众数为 x，y 和 z，说明这三个数字出现的频次是一样多。总计为九个数字，因此可得该集合为： x, y, z, x, y, z, $2x+5$, $2y$ 和 $2z-3$，计算平均数可得 $\frac{x+y+z+x+y+z+2x+5+2y+2z-3}{9} = 20$，整理得 $4(x+y+z)=178$，选 A。

6	D	【题干与问题】：有多少个四位数，满足其百位为偶数，十位为奇数？ 【解析】：任意一个四位数一共有四个数字，其中千位可以取 1～9 中任意一个，一共 9 种可能；百位可以取 0，2，4，6，8，一共 5 种可能；十位可以取 1，3，5，7，9，一共 5 种可能；个位可以取 0～9 任意一个，一共 10 种可能。所以可得一共的可能性为：$9 \times 5 \times 5 \times 10 = 2,250$，选 D。
7	672	【题干与问题】：一个五位数第一位是 9，第三位是 3 或者 6，并且没有数字重复，这样的五位数有几个？ 【解析】：根据题意可得，第一位（万位）是 9，只有一种可能；第三位（百位）可取 3 或者 6，只能有两种可能。由于每次只能取 1 个，又要保证不会出现重复，因此剩下的三个位置可以选择的数字分别有 8 种、7 种和 6 种可能，所以一共有：$1 \times 8 \times 2 \times 7 \times 6 = 672$ 种可能。
8	$\frac{1}{10}$ or 0.1	【题干】：从 1～5 中随机抽取 3 个不相同的数字。 【问题】：抽到 1，2 和 3 的概率是多少？ 【解析】：根据题意可得，从五个数字中选 3 个，没有顺序要求，所以为组合问题，即： $$C_5^3 = \frac{5!}{3!(5-3)!} = \frac{5 \times 4 \times 3!}{3! \times 2!} = 10$$ 因此有 10 种选法，1，2 和 3 是其中的一种可能，所以有 $\frac{1}{10}$ 种可能。答案可写成 $\frac{1}{10}$ 或者 0.1。

6. 乘方与根数

知识点

1. 指数（exponents）：n 个 x 相乘（$x \neq 0$），可以表示成 x^n，在这样的表达式中 x 被称为底（base），n 被称为指数（exponent）。常见的表达式有：

$x^0 = 1$，$x^a \times x^b = x^{a+b}$，$\frac{x^a}{x^b} = x^{a-b}$，$(x^a)^b = x^{a \times b}$，$x^{-a} = \frac{1}{x^a}$，$(x \times y)^a = x^a \times y^a$。

2. 根（roots）：把 $\sqrt[a]{x}$ 称为对 x 开 a 次方，开方之后的值就叫做根（root），常见的表达式有：

$x^{\frac{a}{b}} = (\sqrt[b]{x})^a = \sqrt[b]{x^a}$

$\sqrt{a \times b} = \sqrt{a} \times \sqrt{b}$

$\sqrt{\frac{a}{b}} = \frac{\sqrt{a}}{\sqrt{b}}$（$b \neq 0$）

3. 对数（logarithms）：如果 a 的 x 次方等于 N（$a > 0$，且 a 不等于 1），那么数 x 叫做以 a 为底 N 的对数，记作 $x = \log_a N$。常见的表达式有：

$\log_a x + \log_a y = \log_a(xy)$，$\log_a x - \log_a y = \log_a \frac{x}{y}$，$\log_a x^b = b \times \log_a x$，

$a^{\log_a x} = x$，$\log_x y = \frac{\log_a y}{\log_a x}$（$x > 0$，$y > 0$）。

过关测试

1. If $8^{12} = 2^x$. What is the Value of x?
A. 4   B. 12   C. 24   D. 36

2. $\dfrac{a^2 b^{-6} c^{11} d^{-4}}{a^{-5} b^{-2} c^7 d^9} =$

A. $\dfrac{a^3 b^4}{c^4 d^{13}}$ B. $\dfrac{a^7 b^8}{c^4 d^{13}}$ C. $\dfrac{a^7 c^4}{b^4 d^{13}}$ D. $\dfrac{c^7 d^4}{a^4 b^{13}}$

3. If $\log_5 125 = 3$ and $\log_5 25 = 2$, find $\log_5(125 \times 25)$.

4. $a = ||x|^3 - |x-y|^2 - y + x|$. If $x = -2$ and $y = 5$, then $a =$
A. 47   B. 48   C. 49   D. 50

5. If $(2x^2 + 3x - 5)(x + 2) = ax^3 + bx^2 + cx + d$, then $ac - bd =$
A. 70   B. 71   C. 72   D. 73

6. If $2\sqrt{4n^2} + 7 = 39$, what is the value of n?
A. 8   B. 4   C. 2   D. $2\sqrt{2}$

7. If $x^{-\frac{2}{3}} = \dfrac{1}{36}$, then what does x equal?
A. -6 B. 6   C. 18   D. 216

■答案与解析

1	D	根据题意,$8^{12} = (2^3)^{12} = 2^x$,即 $2^{36} = 2^x$,得到 $x = 36$,选 D。
2	C	根据题意,$\dfrac{a^2 b^{-6} c^{11} d^{-4}}{a^{-5} b^{-2} c^7 d^9} = a^{2-(-5)} b^{-6-(-2)} c^{(11-7)} d^{(-4-9)} = a^7 b^{-4} c^4 d^{-13} = \dfrac{a^7 c^4}{b^4 d^{13}}$,选 C。
3	5	根据题意可得:$\log_5(125 \times 25) = \log_5 125 + \log_5 25 = 3 + 2 = 5$。
4	B	根据题意可得:$|x|^3 = 8$,$|x-y|^2 = 49$,原式 $= |8 - 49 - 5 + (-2)| = 48$,选 B。
5	C	根据题意,将左式展开可得 $2x^3 + 3x^2 - 5x + 4x^2 + 6x - 10 = 2x^3 + 7x^2 + x - 10$,对应右式可得 $a = 2$,$b = 7$,$c = 1$,$d = -10$,因此 $ac - bd = 2 + 70 = 72$,选 C。
6	A	根据题意,$2\sqrt{4n^2} = 32$,因此 $\sqrt{4n^2} = 16$,得到 $n = 8$,选 A。
7	D	根据题意,$\dfrac{1}{x^{\frac{2}{3}}} = \dfrac{1}{36}$,$x^{\frac{2}{3}} = 36$,转换可得 $x = (36)^{\frac{3}{2}} = 216$,选 D。

7. 代数

知识点

1. 常见的代数式有:

$(x - y)(x + y) = x^2 - y^2$

$(x - y)^2 = (x - y)(x - y) = x^2 - 2xy + y^2$

$(x + y)^2 = (x + y)(x + y) = x^2 + 2xy + y^2$

$(x+y)^3 = x^3 + 3x^2y + 3xy^2 + y^3$

$(x-y)^3 = x^3 - 3x^2y + 3xy^2 - y^3$

$x^3 - y^3 = (x-y)(x^2 + xy + y^2)$

$x^3 + y^3 = (x+y)(x^2 - xy + y^2)$

2. 不等式：

$<$：less than

$>$：greater than

$\leq$：less than or equal to

$\geq$：greater than or equal to

不等式的解法与等式相同，但是不等式两边同时乘以或除以相同的负数时，不等号的方向要改变。

带有绝对值的不等式的解法，是把绝对值去掉，拆成两个不等式，分别求解。

3. 余数定理：一个多项式 $f(x)$ 除以一个多项式 $(x-a)$ 的余数是 $f(a)$ 的值。

4. 因式定理：如果 $f(a) = 0$，则 $f(x)$ 必有因式 $x-a$。

5. 多项式的除法：

$$
\begin{array}{r}
ax + b + ae \\
x-e \overline{\smash{)}\, ax^2 + bx + c} \\
-(ax^2 - aex) \\
\hline
(b+ae)x \\
-(bx - be) \\
\hline
aex + be + c \\
-(aex - ae^2) \\
\hline
be + ae^2 + c
\end{array}
$$

因此 $ax^2 + bx + c = (x-e)(ax + b + ae) + be + ae^2 + c$

过关测试

1. For all real numbers v, the operation v^* is defined by the equation $v^* = v - \dfrac{v}{3}$. If $(v^*)^* = 8$, then $v = ?$

A. 15　　　　　　B. 18　　　　　　C. 21　　　　　　D. 24

2. When the polynomial $x^4 - 3x^3 - 7x^2 + 7x + 2$ is divided by $x+2$, the quotient is $x^3 + Bx^2 + Cx + 1$. Find the value of $|B+C|$.

3. The operation & is defined as $r \& s = \dfrac{s^2 - r^2}{r+s}$, where r and s are real numbers and $r \neq -s$. What is the value of $3 \& 4$?

4. The graph of $f(x)$ is shown in the figure (Fig. A1). Which of the following accurately represents the function $f(x)$?

A. $f(x) = at(x+3)(x+4)$

B. $f(x) = (x-3)(x-4)$

C. $f(x) = (x+3)(x-4)$

D. $f(x) = x(x-3)(x-4)$

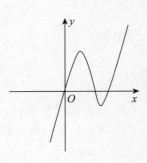

Fig. A1

5. 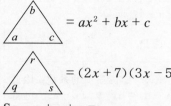 $= x^2 - 5x + 6$, $\boxed{a} = 2$, $a =$
A. -4 B. 4 C. -4 and 4 D. 1 and 4

6. $m \sharp n = m^2 - n^2$, $\dfrac{m \sharp n}{m-n} =$
A. $m \sharp n$ B. $m - n$ C. $m \times n$ D. $m + n$

7.
$\triangle_{a\ c}^{\ b} = ax^2 + bx + c$

$\triangle_{q\ s}^{\ r} = (2x+7)(3x-5)$

So $q + r + s =$
A. -6 B. -10 C. -15 D. -18

8. Which the following represents the statement "When the sum of the squares of $2a$ and $3b$ are added to the difference between $8c$ and $7d$, the result is 3 more than e"?
A. $(2a)^2 + (3b)^2 + (8c - 7d) = e + 3$ B. $(2a)^2 + (3b)^2 + (8c - 7d) + 3 = e$
C. $(2a + 3b)^2 + (8c - 7d) + 3 = e$ D. $(2a + 3b)^2 + (8c - 7d) = e + 3$

9. If $g(f(x)) = \dfrac{5\ln(2^x + 1) - 2}{\ln(2^x + 1) + 3}$ and $g(x) = \dfrac{5x - 2}{x + 3}$, then $f(x) =$
A. $\ln x$ B. $\ln 2x$ C. $\ln(2^x + 1)$ D. $\ln x^2$

■ 答案与解析

1	B	【题干与问题】：对所有实数 v，v^* 定义为计算式 $v^* = v - \dfrac{v}{3}$，如果 $(v^*)^* = 8$，则 v 的值为多少？ 【解析】：根据题意将 v^* 作为一个整体代入 $v^* = v - \dfrac{v}{3}$ 可得 $(v^*)^* = v^* - \dfrac{v^*}{3} = 8$，解得 $v^* = 12$，则 $v = 18$，选 B。
2	2	【题干与问题】：多项式 $x^4 - 3x^3 - 7x^2 + 7x + 2$ 除以 $x+2$，商为 $x^3 + Bx^2 + Cx + 1$。则 $\lvert B+C \rvert$ 的值是多少？ 【解析】：本题采用多项式的整除法： $\begin{array}{r} x^3 - 5x^2 + 3x + 1 \\ x+2 \overline{\smash{\big)}\,x^4 + 3x^3 - 7x^2 + 7x + 2} \\ \underline{-x^4 + 2x^3} \\ -5x^3 - 7x^2 \\ \underline{--5x^3 - 10x^2} \\ 3x^2 + 7x \\ \underline{-3x^2 + 6x} \\ x+2 \\ \underline{-x+2} \\ 0 \end{array}$ 对应上式得 $B = -5$，$C = 3$，因此 $\lvert B+C \rvert = \lvert -5+3 \rvert = 2$。

3	1	【题干与问题】：定义算式 $r\&s = \dfrac{s^2 - r^2}{r + s}$，其中 r 和 s 都是实数且 $r \neq -s$，则 $3\&4$ 的值为多少？ 【解析】：根据题意，将 $r=3$ 和 $s=4$ 代入得 $3\&4 = \dfrac{4^2 - 3^2}{3 + 4} = 1$。
4	D	【题干与问题】：如图所示，下面哪一个式子正确描述了 $f(x)$？ 【解析】：根据图像与 x 轴的交点有三个判断符合这一条件的只能是 D 项。
5	D	【解析】：根据题意可得 $\boxed{a} = a^2 - 5a + 6 = 2$，解得 $a=1$ 和 $a=4$，选 D。
6	D	【解析】：根据题意可得 $\dfrac{m \# n}{m-n} = \dfrac{(m+n)(m-n)}{m-n} = m+n$，选 D。
7	D	【解析】：根据题意可得，第一个三角形的三个角中的数字正好对应右侧三项的系数，因此第二个三角形对应的式子应该是 $qx^2 + rx + s = (2x+7)(3x-5) = 6x^2 + 11x - 35$。可得 $q=6$，$r=11$，$s=-35$，因此 $q+r+s = -18$，选 D。
8	A	【题干与问题】：下面哪一个式子对应该表述？ 【解析】：根据题意，本题有几个分句，需要一一拆分： 1）Squares of $2a$ and $3b$：$2a$ 和 $3b$ 的平方，写成 $(2a)^2$ 和 $(3b)^2$； 2）Sum of those squares：这些平方和，写成 $(2a)^2 + (3b)^2$； 3）Difference between $8c$ and $7d$：$8c$ 和 $7d$ 的差，写成 $(8c - 7d)$； 4）Sum of squares added to the difference：这些平方和与这个差的和，写成 $(2a)^2 + (3b)^2 + (8c - 7d)$； 5）3 more than e：比 e 多 3，写成 $e+3$； 6）Result is 3 more than e：这个结果比 e 多 3，写成 $(2a)^2 + (3b)^2 + (8c-7d) = e+3$，选 A。
9	C	【解析】：根据题意，将 $g(f(x))$ 和 $g(x)$ 对应起来看 $\dfrac{5\ln(2^x+1) - 2}{\ln(2^x+1) + 3} \quad \dfrac{5x - 2}{x + 3}$ 可见 $f(x) = \ln(2^x + 1)$，选 C。

8. 解析几何

知识点

1. 平面直角坐标系：笛卡尔建立了一个平面，在这个平面中有两条相互垂直的直线 x 轴和 y 轴，两条线的交点为原点（origin）。

2. 平面直角坐标系中两个点的坐标为 $A(x_A, y_A)$ 和 $B(x_B, y_B)$，则两个点之间的距离为 $\overline{AB} = \sqrt{(x_A - x_B)^2 + (y_A - y_B)^2}$，将 A 和 B 两点用一条线段相连，则线段 AB 的中点（midpoint）的坐标为 $x = \dfrac{x_A + x_B}{2}$，$y = \dfrac{y_A + y_B}{2}$。

3. 平面直角坐标系中的直线方程：

点斜式：$y - y_0 = k(x - x_0)$，说明该直线过点 (x_0, y_0)，斜率为 k。

截距式：$y = kx + b$，其中 k 是斜率，b 是纵截距。如果该直线过 $A(x_A, y_A)$ 和 $B(x_B, y_B)$，则：

$$k = \frac{y_A - y_B}{x_A - y_B}$$

过关测试

1. Points A, B and C are co-linear. If $\overline{AC} = \overline{BC}$ and $A = (-2, 3)$, $B = (2, -4)$, then $C =$

A. $\left(0, -\dfrac{1}{2}\right)$ B. $(0, -1)$

C. $\left(-\dfrac{1}{2}, 0\right)$ D. $\left(\dfrac{1}{2}, 0\right)$

2. Points $R = (6, 4)$, $Q = (4, 3)$, and $P = (-2, b)$ are co-linear. What is the value of b?

A. 4   B. 3
C. 2   D. 0

3. What is the perimeter of the triangle with vertices $L = (1, 5)$, $M = (-3, -3)$, and $N = (3, 1)$?

A. $6\sqrt{10} + 4\sqrt{13}$ B. $6\sqrt{5} + 2\sqrt{13}$
C. $4\sqrt{5} + 2\sqrt{10}$ D. $8\sqrt{10}$

4. In the figure (Fig. A2), the graph of the function $y = 3x^2 - 8x + 4$ is shown. The function intersects the y-axis at L and the x-axis at N. Line segments LM and MN are perpendicular, and LM is parallel to the x-axis. What is the area of triangle LMN?

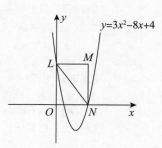

Fig. A2

A. $\dfrac{1}{2}$ B. $\dfrac{3}{8}$
C. 4   D. 8

5. In a rectangular coordinate system, the center of a circle has coordinates $(8, 3)$, and the circle touches the y-axis at only one point. What is the radius of the circle?

6. The vertices of rectangle R are $(0, 0)$, $(0, 4)$, $(5, 0)$, and $(5, 4)$. Let S be the rectangle that consists of all points $(3x, y-2)$ where (x, y) is in R. What is the area of rectangle S?

A. 20   B. 30
C. 40   D. 60

■答案与解析

1	A	【题干】：点 A、B 和 C 共线，且线段 AC 长度与 BC 的长度相等，点 A 的坐标为 $(-2, 3)$，点 B 的坐标为 $(2, -4)$。 【问题】：点 C 的坐标为多少？ 【解析】：根据题意，点 A、B 和 C 共线，且线段 AC 的长度与线段 BC 的长度相等，因此可得三个点之间的关系如图附2所示： 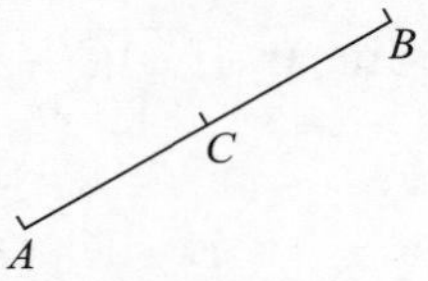 图附2 由于三个点共线，所以点 C 横坐标为 $\frac{-2+2}{2}=0$，点 C 纵坐标为 $\frac{3-4}{2}=-\frac{1}{2}$，选 A。
2	D	【题干】：点 R，Q 和 P 共线，R 的坐标为 $(6, 4)$，Q 的坐标为 $(4, 3)$，P 的坐标为 $(-2, b)$。 【问题】：b 的值为多少？ 【解析】：根据题意，由于三个点共线，所以任意两点相连所成的线段其斜率相等。线段 RQ 的斜率为 $\frac{4-3}{6-4}=\frac{1}{2}$。同理线段 QP 的斜率也为 $\frac{1}{2}$，可得 $\frac{1}{2}=\frac{3-b}{4+2}$，解得 $b=0$，选 D。
3	C	【题干与问题】：顶点为 L，M 和 N 的三角形周长为多少？ 【解析】：根据题意，根据两个点的坐标可以得到两点连线的线段长度：线段 LM 长度为 $2\sqrt{5}$，线段 MN 长度为 $2\sqrt{10}$，线段 LN 长度为 $2\sqrt{5}$，因此△LMN 的周长为 $4\sqrt{5}+2\sqrt{10}$，选 C。
4	C	【题干】：函数 $y=3x^2-8x+4$ 的图像如图所示，该函数的图像与 y 轴的交点为 L，与 x 轴的交点为 N。线段 LM 和 MN 相互垂直，线段 LM 与 x 轴平行。 【问题】：△LMN 的面积为多少？ 【解析】：根据题意，点 L 在函数的图像上，且点 L 是函数图像与 y 轴的交点，可以将 $x=0$ 代入 $y=3x^2-8x+4=4$，因此得到点 L 的坐标为 $(0, 4)$。点 N 是函数图像与 x 轴的交点，因此可以令 $y=3x^2-8x+4=0$，解得 $x_1=\frac{2}{3}$，$x_2=2$。根据图像可得到点 N 的坐标为 $(2, 0)$。这样就可以得到线段 MN 的长度为 4，线段 LM 的长度为 2，△LMN 是一个直角三角形，面积 $=\frac{1}{2}\times LM\times MN=4$，选 C。
5	8	【题干】：在直角坐标系中，一个圆的圆心的坐标为 $(8, 3)$，该圆与 y 轴相交于一个点。 【问题】：该圆的半径是多少？ 【解析】：根据题意，如图附3所示，由于圆与 y 轴相交于一个点，因此圆的半径长度值就是圆心横坐标值，即 $x=8$，因此圆的半径为 8。 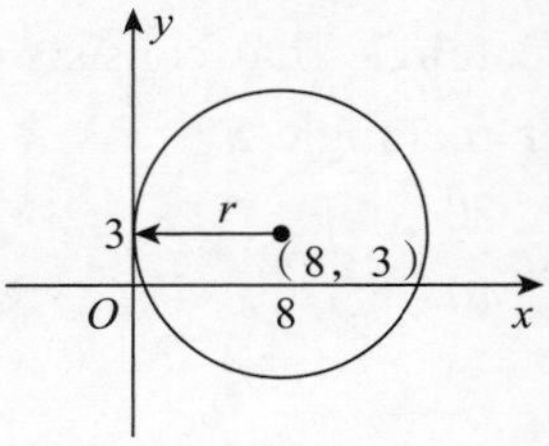 图附3

6	D	【题干】：长方形 R 的四个顶点坐标分别为（0，0），（0，4），（5，0）和（5，4），令长方形 S 包含所有的点（$3x$，$y-2$），而点（x，y）在长方形 R 中。 【问题】：长方形 S 的面积是多少？ 【解析】：根据题意，长方形 R 和 S 的四个点坐标如表附3所示： 表附3 |				
R	（0，0）	（0，4）	（5，0）	（5，4）
S	（0，-2）	（0，2）	（15，-2）	（15，2） | 如图附4所示： 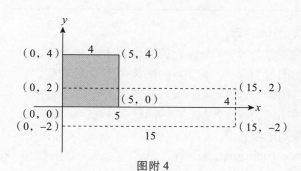 图附4 所以，面积 $S = 15 \times 4 = 60$，选 D。

9. 函数与数学模型

知识点

1. 函数（function）：设在某变化过程中有两个变量 x 和 y，如果 y 随着 x 而变化，当 x 取某个特定的值，y 也会根据确定的关系取相应的值，那么称 y 是 x 的函数，表示为 $y = f(x)$。其中 x 称为自变量（independent variable），y 称为因变量（dependent variable）。x 的变化范围称为函数的"定义域（domain）"，与 x 对应的 y 的取值称为"函数值"，其全体称为函数的"值域（range）"。

2. 函数的图像变换。

1) $y = f(x)$：图像移动见表附4。

表附4

移动形式	移动规则
$y = f(x) \rightarrow y = f(x+a)$	当 $a > 0$ 时，把 $f(x)$ 的图像沿着 x 轴向左平移 a 个单位 当 $a < 0$ 时，把 $f(x)$ 的图像沿着 x 轴向右平移 $-a$ 个单位
$y = f(x) \rightarrow y = f(x)+b$	当 $b > 0$ 时，把 $f(x)$ 的图像沿着 y 轴向上平移 b 个单位 当 $b < 0$ 时，把 $f(x)$ 的图像沿着 y 轴向下平移 $-b$ 个单位

2) $y = f(x)$：图像缩放见表附5。

表附5

缩放形式	缩放规则
$y = f(x) \to y = Af(x)$	当 $A > 1$ 时，横坐标不变，纵坐标伸长到原来的 A 倍 当 $0 < A < 1$ 时，横坐标不变，纵坐标缩短到原来的 $\frac{1}{A}$ 倍
$y = f(x) \to y = f(ax)$	当 $a > 1$ 时，纵坐标不变，横坐标缩短到原来的 $1/a$ 当 $0 < a < 1$ 时，纵坐标不变，横坐标伸长到原来的 a 倍

3) $y = f(x)$ 图像对称。

$y = f(x)$ 的图像与 $y = f(-x)$ 的图像关于 y 轴对称。

$y = f(x)$ 的图像与 $y = -f(x)$ 的图像关于 x 轴对称。

$y = f(x)$ 的图像与 $y = -f(-x)$ 的图像关于原点对称。

$y = f(x)$ 的图像与 $y = f^{-1}(x)$ 的图像关于 $y = x$ 对称。

3. 奇函数和偶函数：

偶函数（even function）满足 $f(-x) = f(x)$，奇函数（odd function）满足 $f(-x) = -f(x)$。

4. 函数的复合。

$f \circ g = f(g(x))$，例如已知 $f(x) = 2x + 3$，$g(x) = x - 3$，则 $f \circ g = f(g(x)) = f(x - 3) = 2(x - 3) + 3 = 2x - 3$，$g \circ f = g(f(x)) = g(2x + 3) = 2x + 3 - 3 = 2x$。

5. 反函数（$f^{-1}(x)$）。

设函数 $y = f(x)$，若找得到一个函数 $g(y)$，在每一处 $g(y)$ 都等于 x，这样的函数 $x = g(y)$ 叫做函数 $y = f(x)$ 的反函数，记作 $y = f^{-1}(x)$，例如：$f(x) = 5x + 2$，则 $f^{-1}(x) = \frac{x - 2}{5}$。

6. 一次函数 $y = ax + b$，其中 a 代表该直线的斜率，b 为直线在 x 轴上的截距，该直线与 x 轴的交点为 $\left(-\frac{b}{a}, 0\right)$，与 y 轴的交点为 $(0, b)$，其中 $a \neq 0$。

7. 二次函数的表达式为：$y = ax^2 + bx + c$，当 $a > 0$ 时，图像的开口朝上；当 $a < 0$ 时，图像的开口朝下。二次函数的表达式可以写成 $y = a(x - h)^2 + k$（$a \neq 0$，a, h, k 为常数），顶点坐标为：(h, k)。当 $a > 0$ 时，$x = h$ 时函数有最小值 k；当 $a < 0$ 时，$x = h$ 时函数有最大值 k。也可以写成 $y = a(x - b)(x - c)$，函数与 x 轴的交点为 $(b, 0)$ 和 $(c, 0)$。

过关测试

1. The graph of $f(x) = x^3 - 2$ is shown in the figure (Fig. A3). Which of the following choice is the graph of $g(x) = (x + 3)^3 - 6$?

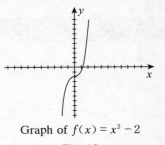

Graph of $f(x) = x^3 - 2$

Fig. A3

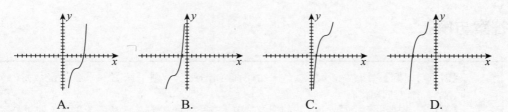

A.   B.   C.   D.

2. A periodic function is a function that repeats a pattern of range values forever. Two cycles of periodic function f are shown in the graph (Fig. A4) of $y = f(x)$. What is the value of $f(89)$?

A. -2 B. -1 C. 0   D. 1

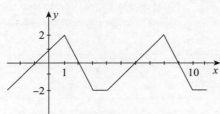

Fig. A4

3. $f(x) = 3x^2 - 7$ and $g(x) = 2x^3 - 4x + 2$. What is the value of $g[f(2)]$?

4. The cost of operating a Frisbee company in the first year is $\$10,000$ plus $\$2$ for each Frisbee. Assuming the company sells every Frisbee it makes in the first year for $\$7$, how many Frisbees must the company sell to break even?

A. 1,000   B. 1,500   C. 2,000   D. 2,500

5. The water level in a bay changes with the tides. The tides go through a full cycle every 12 hours, with one low and one high tide. Which of the following graphs shows the water level in the bay during a 24-hour period starting with the high tide?

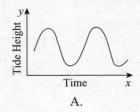 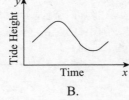 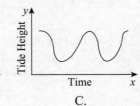 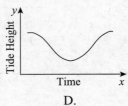

A.   B.   C.   D.

6. A system of three equations and their graphs in the xy-plane are shown in the picture (Fig. A5). How many solutions does the system have?

$$x^2 + y^2 = 10$$
$$y = 4 - x^2$$
$$y = 2 - x$$

A. One   B. Two
C. Three   D. Four

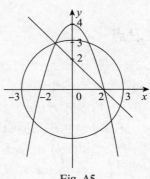

Fig. A5

■ 答案与解析

1	B	【题干与问题】：函数 $f(x)=x^3-2$ 的图像如图所示，则下面哪一个图是函数 $g(x)=(x+3)^3-6$ 的图像？ 【解析】：根据题意可得：$g(x)$ 的图像是由 $f(x)$ 的图像向左移动三个单位，向下移动四个单位得到。满足该条件的图为 B 项的图，选 B。
2	B	【题干】：周期函数是指一个函数在取值范围内呈现周期性不断重复，函数 $y=f(x)$ 的两次周期性重复如图所示。 【问题】：$f(89)$ 的值是多少？ 【解析】：根据题意可得，该图像的两个峰值分别对应的是 $x=1$ 和 $x=8$，可见该周期函数图像是在 x 轴上每 $8-1=7$ 个单位呈现周期性重复。89 除以 7 商为 12，余数为 5，说明 $f(89)=f(5)=-1$，选 B。
3	232	【题干与问题】：已知 $f(x)=3x^2-7$ 和 $g(x)=2x^3-4x+2$，则 $g[f(2)]$ 的值是多少？ 【解析】：根据题意可得：$f(2)=3\times2^2-7=5$，$g(5)=2\times5^3-4\times5+2=232$。
4	C	【题干】：运营一家飞盘公司第一年的成本为 \$10,000，加上每卖出一个飞盘成本为 \$2。假设第一年该公司每个飞盘卖 \$7。 【问题】：该公司需要卖出多少飞盘使得第一年不赔不赚？ 【解析】：根据题意可得，不赔不赚指的是收入与成本相等，设需要卖出飞盘的数量为 f，可得关系式：$10,000+2\times f=7\times f$，解得 $f=2,000$，选 C。
5	C	【题干】：一个港湾的水位随着潮汐有一定的变化，每 12 小时有一次完整的潮汐，包括一次涨潮和落潮。 【问题】：下面哪一个图像代表的是该港湾水位在一个 24 小时内，并且是以高水位开始的变化趋势？ 【解析】：根据题意，24 小时内该港湾的水位经历了两次完整的变化过程，且是以涨潮开始的，满足这两个条件的图只有 C 项。
6	A	【题干与问题】：一个方程组中三个方程及其图像如图所示，则该方程组有几个解？ 【解析】：根据题意，方程组的解即是方程对应图像的交点，从图中可以看出三个方程的图像只有一个公共点，因此该方程组有一个解，选 A。

10. 平面几何

知识点

1. 角（angle）：0°到 90°之间的角为锐角（acute angle），90°角为直角（right angle），90°到 180°之间的角为钝角（obtuse angle）。

2. 三角形（triangle）：三角形的性质如下。

边（sides）：三角形的任意两边之和大于第三边，任意两边之差小于第三边。

内角（interior angle）：所有三角形都有一个共同的性质，即三个内角之和为 180°。

周长（perimeter）：三角形的周长等于三边之和。

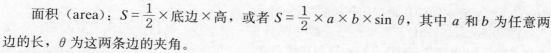

面积（area）：$S = \frac{1}{2} \times 底边 \times 高$，或者 $S = \frac{1}{2} \times a \times b \times \sin\theta$，其中 a 和 b 为任意两边的长，θ 为这两条边的夹角。

特殊三角形：

1) 等腰三角形（isosceles triangle）：如果三角形有两条边相等，则该三角形为等腰三角形。如果沿着顶点作底边的垂线，交点刚好平分底边。

2) 等边三角形（equilateral triangle）：如果三角形的三条边都相等，则该三角形为等边三角形。等边三角形的三个内角也相等，为 60°。如果设 a 为边长，则周长为 $3a$，面积为 $\frac{\sqrt{3}}{4}a^2$。

3) 直角三角形（right triangle）：如果三角形的一个内角为 90°，则该三角形为直角三角形。直角的两条邻边为直角边（leg），对边为斜边（hypotenuse）。

三角形的勾股定理：设 a 和 b 为直角边，c 为斜边，则满足 $a^2 + b^2 = c^2$。请注意两个特殊的勾股定理：一个是 30°- 60°- 90°三角形，边长满足 $1:\sqrt{3}:2$，一个是 45°- 45°- 90°三角形，边长满足 $1:1:\sqrt{2}$。

相似三角形（similar triangle）：如果两个三角形的三个角对应分别相等，只是边长不相等，则称这两个三角形为相似三角形。两个相似三角形的对应角对应相等，且两个相似三角形的边长对应成比例。如果两个三角形相似，边的相似比为 a，则两者的面积相似比为 a^2。如果两个三角形完全相同，则称为全等三角形（congruent triangle）。

3. 四边形（quadrilateral）：要掌握常见四边形的名称、周长与面积的计算方法，以及一些特殊的性质。

1) 正方形（square）：四条边都相等，且内角都为直角的四边形。设正方形的边长为 a，则正方形的周长为 $4a$，面积为 a^2，对角线长为 $\sqrt{2}a$。

2) 长方形（rectangle）：内角都是直角的四边形。设长方形的长（length）为 a，宽（width）为 b，则周长为 $2(a+b)$，面积为 $a \times b$，对角线（diagonal）为 $\sqrt{a^2+b^2}$。

3) 平行四边形（parallelogram）：一个四边形对边两两平行，则该四边形为平行四边形。正方形和长方形都是特殊的平行四边形。设平行四边形的边长分别为 a 和 b，则周长为 $2(a+b)$，面积为 $a \times b \times \sin\theta$，其中 θ 为 a 和 b 的夹角。

4) 菱形（rhombuse）：一个四条边都相等的四边形。菱形就是 $a = b$ 的平行四边形，所以周长和面积的算法同上。菱形有一个值得注意的性质就是两条对角线是相互垂直的。

5) 梯形（trapezoid）：只有一组对边平行的四边形为梯形，其面积为：$\frac{a+b}{2} \times h$，其中 a 和 b 分别是上底和下底的长度，h 是高。

4. 多边形（polygon）：主要考查的是正多边形（regular polygon）。多边形的内角和为 $(n-2) \times 180°$，其中 n 为边的数目。多边形的面积算法一般是把多边形切割成几个规则的形状。

5. 圆（circle）：在平面内，绕着一个点等距的点的集合，构成圆。中心的这个点为圆心（center），等距的距离为半径（radius，r），圆上任意两点之间的连线为弦（chord），最长的弦是直径（diameter，d），直径刚好是半径的两倍，周长为 $2\pi r$ 或者为 πd，面积

为 πr^2。注意有关圆的几个特殊几何图形：

1) 切线（tangent）：如果一条直线与圆只有一个交点，则这条线称为圆的切线，这个交点称为切点，圆心到切点的连线与切线垂直。

2) 扇形（sector）：圆的两条半径可以把圆割成一个封闭的面积，其中两条半径的夹角为圆心角（central angle），一般用 θ 来表示，则扇形的弧长（arc）可以表示为 arc = $\theta \times r$，面积为 $S = \frac{1}{2}\theta r^2$，$r$ 为半径。注意这里的 θ 需要转换为弧度制。

6. 平行线。一条直线与两条平行线相交会形成若干个角，如图附5所示，$\angle A$ 与 $\angle B$ 为对顶角（vertical angle），$\angle B$ 与 $\angle C$ 为内错角（alternate angle），$\angle A$ 与 $\angle C$ 为同位角（corresponding angle）。

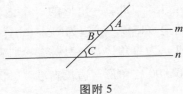

图附5

过关测试

1. $\triangle ABC$ is an equilateral triangle and $\triangle ADC$ is an isosceles triangle (Fig. A6). If $AD = 12$, what is the area of the shaded region?

A. $108\sqrt{3}$ B. $72\sqrt{3}$

C. $36\sqrt{3}$ D. 144

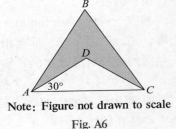

Note: Figure not drawn to scale

Fig. A6

2. Many cities try to work "greenspace" into their city planning because living plants help filter the city's air, reducing the effects of pollution. The figure (Fig. A7) shows the plans for a new greenspace around City Hall, which will be created by converting portions of the existing parking lots. If the width of each parking lot is the same as the width of the City Hall building, how many thousands of square feet of greenspace will there be after the conversion? Round to the nearest thousand and enter your answer in terms of thousands. (For example, enter 14,000 as 14.)

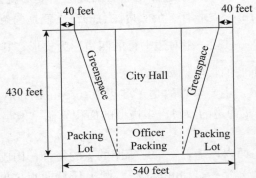

Note: Figure not drawn to scale

Fig. A7

3. An equilateral triangle is inscribed in a circle with a radius of 2 meters. What is the area of this triangle?
A. $\sqrt{3}$ B. $2\sqrt{3}$ C. $3\sqrt{3}$ D. $\pi\sqrt{3}$

4. The angles of a pentagon are in ratio 9 : 10 : 12 : 14 : 15. What is the sum of measures of the smallest and largest angles?
A. 54°   B. 81°   C. 135°   D. 216°

5. In the figure (Fig. A8), P is the center of a circle and AC is its diameter. What is the value of x?
A. 60   B. 50   C. 40   D. 30

6. What is the area of the shaded region (Fig. A9)?
A. 24π B. 60π C. 90π D. 120π

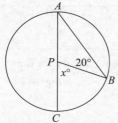

Note: Figure not drawn to scale
Fig. A8

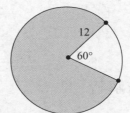

Note: Figure not drawn to scale
Fig. A9

7. In the figure (Fig. A10), the circles' centers at N and O intersect at X and Y, and points N, X, Y, and O are co-linear. The lengths of $\overline{MN}$, $\overline{OP}$ and $\overline{XY}$ are 10, 8, and 3 inches, respectively. What is the length, in inches, of NO?

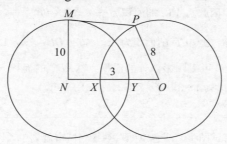

Note: Figure not drawn to scale
Fig. A10

8. In the figure (Fig. A11), $\overline{AB} \parallel \overline{CD}$. What is the value of y?
A. 110°   B. 120°   C. 135°   D. 140°

9. The figure (Fig. A12) shows two concentric circles with center O, If $OD = 3$, $DB = 5$, and the length of arc AB is 5π, What is the length of arc CD?
A. $\dfrac{7}{4}\pi$ B. $\dfrac{15}{8}\pi$ C. 3π D. $\dfrac{25}{8}\pi$

SAT-1 数学轻松突破 800 分：思路与技巧的飞跃

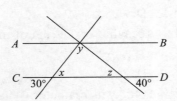

Note：Figure not drawn to scale

Fig. A11

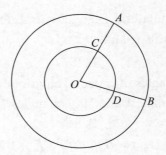

Note：Figure not drawn to scale

Fig. A12

10. In the figure（Fig. A13），lines m and n are parallel. What is the value of z ?

A. 65°
B. 115°
C. 125°
D. 145°

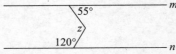

Note：Figure not drawn to scale

Fig. A13

■答案与解析

1	B	【题干】：△ABC 是一个等边三角形，△ADC 是一个等腰三角形，AD 长为 12。 【问题】：图中阴影部分的面积是多少？ 【解析】：根据题意，从点 B 向线段 AC 作一条垂线交 AC 于点 E，如图附 6 所示：根据三角形的性质可得，这条垂线也过点 D，且△ADE 是一个 30°-60°-90°的直角三角形。可得 DE 长度为 6，AE 长度为 $6\sqrt{3}$。同理△ABE 也是一个 30°-60°-90°的直角三角形，可得 BE 长度为 18。因此△ABC 的面积为 $\frac{1}{2} \times AC \times BE = \frac{1}{2} \times 12\sqrt{3} \times 18 = 108\sqrt{3}$，△ACD 的面积为 $\frac{1}{2} \times AC \times DE = \frac{1}{2} \times 12\sqrt{3} \times 6 = 36\sqrt{3}$。最终可得阴影部分的面积 = △ABC 的面积 - △ACD 的面积 = $108\sqrt{3} - 36\sqrt{3} = 72\sqrt{3}$，选 B。	图附 6
2	60	【题干】：很多城市试图将"绿色空间"纳入到城市规划中，因为活的植物具有帮助过滤城市空气、减少污染的效应。下图显示出市政厅周围的绿色空间设计，主要是将现有的停车场改造为绿色空间。如果停车场的宽度与市政厅建筑的宽度一致。 【问题】：改造后将有多少千平方英尺的绿色空间？（将最终的答案保留到千位，并且以千位为基本单位，例如 14,000 为一万四千，以千位为单位即 14） 【解析】：根据题意，由于停车场的宽度与市政厅建筑的宽度一致，说明停车场与市政厅建筑的宽度均为 180 英尺。而绿色空间为一个直角三角形，高为 430 英尺，底为 140 英尺。如图附 7 所示：	

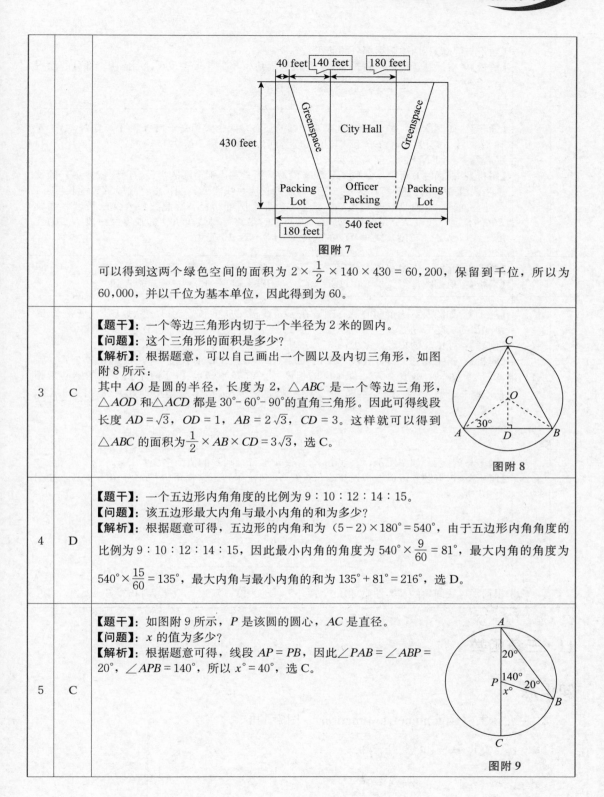

图附7

可以得到这两个绿色空间的面积为 $2 \times \frac{1}{2} \times 140 \times 430 = 60,200$，保留到千位，所以为 60,000，并以千位为基本单位，因此得到为 60。

3	C	【题干】：一个等边三角形内切于一个半径为2米的圆内。 【问题】：这个三角形的面积是多少？ 【解析】：根据题意，可以自己画出一个圆以及内切三角形，如图附8所示： 其中 AO 是圆的半径，长度为2，$\triangle ABC$ 是一个等边三角形，$\triangle AOD$ 和 $\triangle ACD$ 都是 30°-60°-90° 的直角三角形。因此可得线段长度 $AD = \sqrt{3}$，$OD = 1$，$AB = 2\sqrt{3}$，$CD = 3$。这样就可以得到 $\triangle ABC$ 的面积为 $\frac{1}{2} \times AB \times CD = 3\sqrt{3}$，选 C。

图附8

4	D	【题干】：一个五边形内角角度的比例为 9：10：12：14：15。 【问题】：该五边形最大内角与最小内角的和为多少？ 【解析】：根据题意可得，五边形的内角和为 $(5-2) \times 180° = 540°$，由于五边形内角角度的比例为 9：10：12：14：15，因此最小内角的角度为 $540° \times \frac{9}{60} = 81°$，最大内角的角度为 $540° \times \frac{15}{60} = 135°$，最大内角与最小内角的和为 $135° + 81° = 216°$，选 D。

5	C	【题干】：如图附9所示，P 是该圆的圆心，AC 是直径。 【问题】：x 的值为多少？ 【解析】：根据题意可得，线段 $AP = PB$，因此 $\angle PAB = \angle ABP = 20°$，$\angle APB = 140°$，所以 $x° = 40°$，选 C。

图附9

6	D	【题干与问题】：图中阴影部分的面积是多少？ 【解析】：根据题意可得，阴影部分对应的圆周角为 $360°-60°=300°$，因此阴影部分的面积为：$\frac{300}{360}\times\pi\times r^2=\frac{5}{6}\times\pi\times 144=120\pi$，选 D。
7	15	【题干】：如图所示，N 和 O 分别是两个圆的圆心，两个圆相交于点 X 和 Y，点 N，X，Y 和 O 共线。线段 MN、OP 和 XY 的长度分别为 10，8 和 3 英寸。 【问题】：线段 NO 的长度是多少英寸？ 【解析】：根据题意可得，线段 NY 和线段 MN 都是左边那个圆的半径，因此线段 NY 的长度 = 线段 MN 的长度 = 10 英寸，线段 NX 的长度 = 线段 NY 的长度 − 线段 XY 的长度 = 7 英寸。同理，线段 XO 与线段 PO 都是右边那个圆的半径，因此线段 YO 的长度 = 线段 PO 的长度 − 线段 XY 的长度 = $8-3=5$ 英寸。最终可得线段 NO 的长度 = 线段 NX 的长度 + 线段 XY 的长度 + 线段 YO 的长度 = $7+3+5=15$ 英寸。
8	A	【题干与问题】：直线 AB 与 CD 平行，y 值为多少？ 【解析】：根据图可得，$x=30°$，$z=40°$，$y=180°-30°-40°=110°$，选 A。
9	B	【题干】：如图所示两个同心圆，圆心在 O，线段 OD 的长度为 3，线段 DB 的长度为 5，弧 AB 的长度为 5π。 【问题】：弧 CD 的长度为多少？ 【解析】：根据题意可得，线段 OD 的长度 = 3，线段 OB 的长度 = 8，根据弧长公式：弧 AB 的长度 = $\angle AOB \times OB$，弧 CD 的长度 = $\angle COD \times OD$，由于两个弧对应的圆心角相等，因此可得：$\frac{弧 AB 的长度}{线段 OB 的长度}=\frac{弧 CD 的长度}{线段 OD 的长度}$，即 $\frac{5\pi}{8}=\frac{弧 CD 的长度}{3}$，所以弧 CD 的长度 = $\frac{15}{8}\pi$，选 B。
10	B	【题干与问题】：如图所示，直线 m 与直线 n 平行，则 z 的值为多少？ 【解析】：根据题意可以做一条直线与直线 m 和 n 平行，这样可以把 z 分为 z_1+z_2，如图附 10 所示： 图附 10 根据平行线间角的基本性质可知：$z_1=55°$，$z_2=60°$，所以 $z_1+z_2=115°$，选 B。

11. 三角函数

知识点

1. 三角函数（trigonometric function）（图附 11）

Sine of angle $A = \sin A = \dfrac{BC}{AB}$

Cosine of angle $A = \cos A = \dfrac{AC}{AB}$

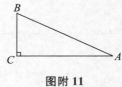

图附 11

Tangent of angle $A = \tan A = \dfrac{BC}{AC}$

Cotangent of angle $A = \cot A = \dfrac{AC}{BC}$

Secant of angle $A = \sec A = \dfrac{AB}{AC}$

Cosecant of angle $A = \csc A = \dfrac{AB}{BC}$

2. 三角恒等式

$\sin^2 A + \cos^2 B = 1$

$\sin 2A = 2\sin A\cos A$

$\cos 2A = 1 - 2\sin^2 A = 2\cos^2 A - 1 = \cos^2 A - \sin^2 A$

$1 + \tan^2 A = \sec^2 A$

$1 + \cot^2 A = \csc^2 A$

$\sin(A + B) = \sin A\cos B + \cos A\sin B$

$\sin(A - B) = \sin A\cos B - \cos A\sin B$

$\cos(A + B) = \cos A\cos B - \sin A\sin B$

$\cos(A - B) = \cos A\cos B + \sin A\sin B$

3. 解三角形（solving triangle）

$\dfrac{A}{\sin A} = \dfrac{B}{\sin B} = \dfrac{C}{\sin C}$

$A^2 = B^2 + C^2 - 2BC\cos A$

$B^2 = A^2 + C^2 - 2AC\cos B$

$C^2 = A^2 + B^2 - 2AB\cos C$

4. 函数图像（表附6）

表附6

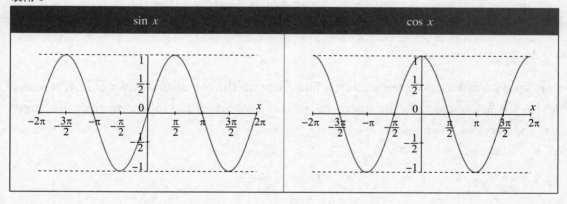

续前表

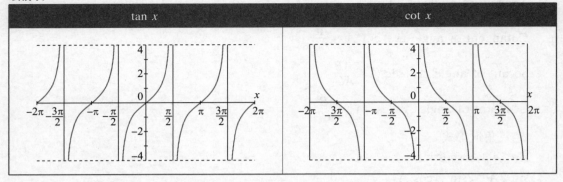

5. 诱导公式（图附 12）

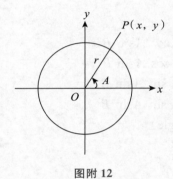

图附 12

$r = 1$

$\sin A = y$，$\cos A = x$，$\tan A = \dfrac{y}{x}$

$\sin(\pi + A) = -\sin A$，$\cos(\pi + A) = -\cos A$，$\tan(\pi + A) = \tan A$

$\sin(-A) = -\sin A$，$\cos(-A) = \cos A$，$\tan(-A) = -\tan A$

$\sin(\pi - A) = \sin A$，$\cos(\pi - A) = -\cos A$，$\tan(\pi - A) = -\tan A$

过关测试

1. A straight ladder is leaned against a house so that the top of the ladder is 12 feet above level ground, as shown in the figure (Fig. A14). Which of the following gives the length (x), in feet, of the ladder?

A. $x = 12\cos 65°$ 　　　　　　　B. $x = 12\sin 65°$

C. $x = \dfrac{12}{\cos 65°}$ 　　　　　　　D. $x = \dfrac{12}{\sin 65°}$

2. For the polygon (Fig. A15), which of the following represents the length, in inches, of FK?

A. 10 　　　　　　　　　　　　B. 30

C. $\dfrac{10}{\sin 70°}$ 　　　　　　　　D. $\dfrac{30}{\sin 70°}$

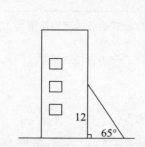

Note: Figure not drawn to scale

Fig. A14

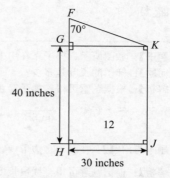

Note: Figure not drawn to scale

Fig. A15

3. A dog, a cat, and a mouse are all sitting in a room. Their relative positions to each other are described in the figure (Fig. A16). Which of the following gives the distance, in feet, from the cat to the mouse?

A. 5.5   B. 5.9   C. 8.57   D. 10.9

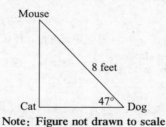

Note: Figure not drawn to scale

Fig. A16

4. $\tan^2 x + 1 =$

A. $\cot^2 x$ B. $\csc^2 x$ C. $\sec^2 x$ D. $\cos^2 x$

5. If $0 < x < 2\pi$ and $5\cos x = \sqrt{5}$, what is the value of $\left(\dfrac{\sin x}{3}\right)^2$.

6. The equation of line M shown (Fig. A17) is $y = -\dfrac{3}{4}x +$

5. Given that angle A is the acute angle formed by the intersection of line M and the y-axis, which expression could be used to find the measure of angle A?

A. $\cos A = \dfrac{3}{4}$

B. $\sin A = \dfrac{4}{3}$

C. $\tan A = \dfrac{4}{3}$

D. $\cos A = \dfrac{4}{5}$

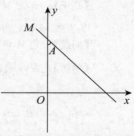

Note: Figure not drawn to scale

Fig. A17

■答案与解析

1	D	【题干】：如图所示，梯子靠在房子的墙上，梯子的顶端到地面的垂直高度为12英尺。 【问题】：梯子的长度是多少？ 【解析】：根据题意可得，梯子靠在房子的墙上，构成了一个直角三角形，因此：$\sin 65° = \dfrac{12}{x}$，解得 $x = \dfrac{12}{\sin 65°}$，选 D。
2	D	【题干与问题】：在图中多边形内线段 FK 的长度是多少？ 【解析】：根据题意可得，线段长度 $GK = HJ = 30$，因此 $\sin 70° = \dfrac{GK}{FK}$，解得 $FK = \dfrac{GK}{\sin 70°} = \dfrac{30}{\sin 70°}$，选 D。
3	B	【题干】：一条狗、一只猫和一只小老鼠都坐在一间房间里，它们的相对位置如图所示。 【问题】：猫到小老鼠的距离是多少？ 【解析】：根据题意，在此三角形中，已知猫和狗的距离为8英尺（斜边），设猫和小老鼠的距离为 x，则满足 $\sin 47° = \dfrac{x}{8}$，解得 $x = 8 \times \sin 47° \approx 5.9$，选 B。
4	C	【解析】：因为 $\cos^2 x + \sin^2 x = 1$，根据题意可得，原式可变形： $\dfrac{\sin^2 x}{\cos^2 x} + \dfrac{\cos^2 x}{\cos^2 x} = \dfrac{\cos^2 x + \sin^2 x}{\cos^2 x} = \dfrac{1}{\cos^2 x} = \sec^2 x$，选 C。
5	$\dfrac{4}{45}$	【题干与问题】：如果 $0 < x < 2\pi$ 并且 $5\cos x = \sqrt{5}$，$\left(\dfrac{\sin x}{3}\right)^2$ 的值是多少？ 【解析】：根据题意可得：$\cos x = \dfrac{\sqrt{5}}{5}$，$\cos^2 x + \sin^2 x = 1$，因此 $\left(\dfrac{\sin x}{3}\right)^2 = \dfrac{1 - \cos^2 x}{9} = \dfrac{1 - \left(\dfrac{\sqrt{5}}{5}\right)^2}{9} = \dfrac{4}{45}$。
6	C	【题干】：直线 M 的方程为 $y = -\dfrac{3}{4}x + 5$，如图所示。角 A 是直线 M 和 y 轴相交所形成的锐角。 【问题】：角 A 可以用以下哪个式子来表示？ 【解析】：根据题意可得，直线 M 的斜率是 $-\dfrac{3}{4}$，可以表示为 $\Delta y = 3$，$\Delta x = 4$，如图附13，角 A 对应的两个直角边为4和3，因此 $\tan A = \dfrac{4}{3}$，选 C。 图附13

12. 立体几何

知识点

立体几何中涉及三维立体图形的面积以及表面积的计算，有正方体（cube），长方体

（rectangular solid），棱柱（prism），圆柱（circular cylinder），圆锥（cone），棱锥（pyramid）和球（sphere）。涉及立体几何问题的计算公式都会列在数学试题部分的第一页，需要大家掌握。

过关测试

1. In the figure（Fig. A18），a rectangular container with the dimensions 10 inches by 15 inches by 20 inches is to be filled with water, using a cylindrical cup whose radius is 2 inches and whose height is 5 inches. What is the maximum number of full cups of water that can be placed into the container without the water overflowing the container?

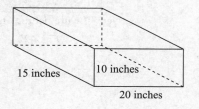 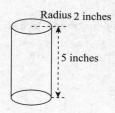

Fig. A18

2. A rectangular prism has dimensions $h = 2$ feet, $w = 7$ feet, and $l = 4$ feet. The prism is cut into two separate rectangular prisms by a plane parallel to one of the faces. What is the maximum increase in the surface area between the original prism and the two separate prisms?
A. 8 ft²   B. 14 ft²   C. 28 ft²   D. 56 ft²

3. The conical tank（Fig. A19）is filled with liquid. How much liquid has poured from the tip of the cone if the water level is 9 feet from the tip?
A. 67π ft³   B. 47π ft³
C. 37π ft³   D. 27π ft³

4. The figure（Fig. A20）is the side view of a pool that is 10 feet wide. What is the volume of the water in the pool when it is filled to 2 feet below the top?
A. 1,120   B. 720   C. 360   D. 200

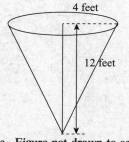

Note: Figure not drawn to scale
Fig. A19

Note: Figure not drawn to scale
Fig. A20

5. If the radius of a circular cylinder（Fig. A21）is decreased by 50%, and its height is simultaneously increased by 60%, what is the change in volume?
A. An increase of 40%   B. A decrease of 40%
C. An increase of 60%   D. A decrease of 60%

6. The figure (Fig. A22) shows a metal square nut with two square faces and a thickness of 1 cm. The length of each side of a square face is 4 cm. A hole with a diameter of 2 cm is drilled through the nut. The density of the metal is 7.9 grams per cubic cm. What is the mass of this nut, to the nearest gram? (Density is mass divided by volume.)

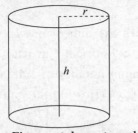

Note: Figure not drawn to scale

Fig. A21

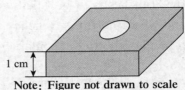

Note: Figure not drawn to scale

Fig. A22

■ 答案与解析

1	47	【题干】：如图所示，一个矩形容器尺寸为 10 英寸×15 英寸×20 英寸，里面可以充满水。现在还有一个圆柱形的杯子，底面半径为 2 英寸，高为 5 英寸。 【问题】：将满杯的水倒入到这个矩形容器中，最多可以倒入多少杯水使得该矩形容器中的水不会溢出？ 【解析】：根据题意，首先可以将矩形容器中可以盛满水的体积算出： $Vrc = 15 \times 20 \times 10 = 3,000$，$Vc = \pi r^2 h = \pi \times 2^2 \times 5 = 20\pi$，因此可以得到需要的杯数：$\frac{3,000}{20\pi} = 47.7$。因为要保证不能有水溢出，所以最后的值≈47。
2	D	【题干】：一个矩形的棱柱尺寸为 2 英尺（高）×7 英尺（宽）×4 英尺（长）。现在有一个平面平行于该棱柱的某一面将该棱柱分成两个棱柱。 【问题】：相对于原始棱柱，形成的新棱柱可以增加的表面积最大是多少？ 【解析】：根据题意，形成的新棱柱增加的表面积就是该平面切过后所形成的两个面，如图附 14 所示： 切之前　　　　　　　切之后 图附 14 因此要想使得新增的表面积最大，只要沿着该棱柱最大的面切割即可得到。因此该面为 7 英尺×4 英尺＝28 平方英尺。由于是增加了两个面，所以最终增加的表面积是 28×2＝56 平方英尺，选 D。

3	C	**【题干】**：如图所示，一个锥形水罐中充满了液体。 **【问题】**：如果要使液面离水罐的尖端为 9 英尺，需要倒出多少液体？ 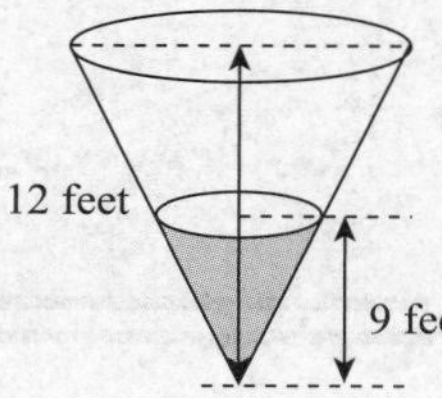 **【解析】**：如图附 15 所示，要满足液面离水罐尖端为 9 英尺，需要倒出液体的体积可以看做是两个锥形的体积差：V_1（图中大体积锥形）$- V_2$（图中小体积锥形）。 根据锥形体积公式可得：$V_1 = \frac{1}{3} \times \pi \times 4^2 \times 12$。 大体积锥形与小体积锥形底面的半径之比与高之比相同，即 $\frac{12}{9} = \frac{4}{r_2}$，解得 $r_2 = 3$， 所以可得 $V_2 = \frac{1}{3} \times \pi \times 3^2 \times 9$，$V_1 - V_2 = 64\pi - 27\pi = 37\pi$，选 C。 图附 15
4	B	**【题干】**：下图展示的是一个宽为 10 英尺水池的侧面视角。 **【问题】**：当这个水池中的水面离顶部还有 2 英尺的时候水池中水的体积是多少？ **【解析】**：根据题意，该水池的侧面为一个矩形加一个梯形，如图附 16 所示： 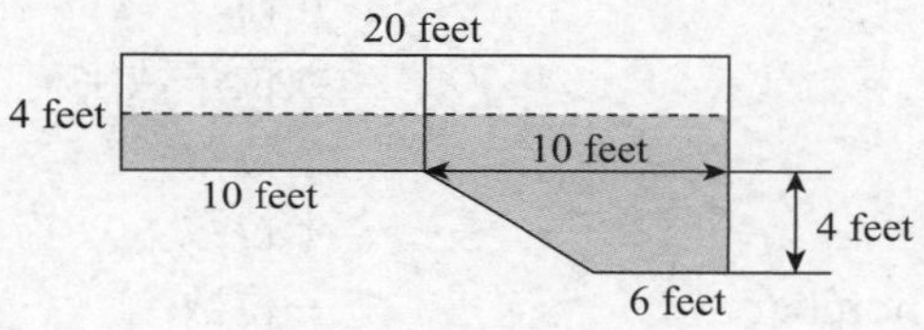 图附 16 其中矩形的面积为 $S_1 = 20 \times 2 = 40$ 平方英尺，梯形的面积为 $S_2 = \frac{1}{2} \times (6 + 10) \times 4 = 32$ 平方英尺。因此侧面积为 $40 + 32 = 72$ 平方英尺，整个体积为 $V = 72 \times 10 = 720$ 立方英尺，选 B。
5	D	**【题干】**：如果圆柱体的底面半径减少 50%，同时高增加 60%。 **【问题】**：则其体积的变化为多少？ **【解析】**：根据题意，设原来的圆柱体的底面半径为 r，高为 h，体积为 $V_1 = \pi \times r^2 \times h$，变化后底面半径为 $\frac{1}{2}r$，h 为 $1.6h$，因此体积为 $V_2 = \pi \times \left(\frac{1}{2}r\right)^2 \times 1.6h$，$\frac{V_2}{V_1} = \frac{\pi \times \left(\frac{1}{2}r\right)^2 \times 1.6h}{\pi \times r^2 \times h} = 0.4$，可见变化后的圆柱体体积是原圆柱体体积的 0.4，说明是减少了 60%，选 D。
6	102	**【题干】**：如图所示，显示的是一个金属方形螺母，含有两个方形的表面，厚度为 1 厘米，方形表面边长为 4 厘米。有一个 2 厘米直径的孔穿过该螺母。该螺母金属的密度为 7.9 克/立方厘米（密度＝质量÷体积）。 **【问题】**：该螺母的质量为多少？（结果保留到整数） **【解析】**：首先计算出该方形螺母的总体积，$V_1 = 4 \times 4 \times 1 = 16$；其次计算中心圆柱体的 $V_2 = \pi r^2 h$；因此可得扣除该圆柱体的体积后该螺母的实际体积 $V = V_1 - V_2 = 16 - \pi \approx 12.86 \text{ cm}^3$。最后计算质量为 $7.9 \times 12.86 = 101.59 \approx 102$。

附录 2 常见数学表达

1. 基本概念和词组

1	operation	运算
2	algebra	代数
3	arithmetic	算数
4	geometry	几何
5	digit	数位；数字
6	units digit	个位数
7	tens digit	十位数
8	single-digit number	一位数
9	two-digit number	两位数
10	equal	相等；等于
11	be equivalent/equal to	与……相等
12	expression	表达式
13	equation	方程式；等式；解析式
14	solution	（方程的）解
15	constant	常量；恒量
16	variable	变量
17	represent/express	（变量、函数等）表示
18	y in terms of x	用 x 表示 y
19	value	值
20	absolute value	绝对值
21	all values	全部的值
22	possible value	可能的值
23	least possible value	最小可能的值
24	maximum	最大的
25	minimum	最小的
26	be closest to	最接近
27	approximate	估计；接近
28	estimation	估计；近似
29	certain	某个

30	overlap	重叠
31	nonoverlapping	不重叠的
32	(anti) clocking	(逆)顺时针方向
33	number	(实数)数目
34	real number, rational number	实数
35	complex number	复数
36	prime	质数/素数
37	greatest common factor, GCF	最大公约数
38	least common multiple, LCM	最小公倍数

2. 运算

1	addition	加法
2	subtraction	减法
3	multiplication	乘法
4	division	除法
5	add/plus	加
6	be added to, be increased by	加;增加
7	exceed…by…	超过……
8	sum	和
9	a total of	总共
10	combination	(两者的)和
11	subtract, minus	减
12	be subtracted from	从……减去
13	difference	差
14	differ by	相差……
15	multiply, time	乘
16	product	积
17	twice, two times	两倍
18	three times	三倍
19	half of	……的一半
20	divide	除
21	be divided by	被……除
22	divisible	可以整除的
23	divisor	除数;因子
24	dividend	被除数
25	quotient	商
26	remainder	余数
27	factor	因子
28	factorization	因式分解

29	factorial	阶乘
30	exponent	指数；幂
31	base	底数
32	square	平方
33	perfect square	完全平方
34	power	乘方
35	A to Bth power	A 的 B 次方
36	root	根
37	radical	根式
38	square root	平方根
39	cube root	立方根
40	radical sign, root sign	根号
41	parentheses	括号
42	distributive law	分配定律
43	algebraic term	代数项
44	algebraic fraction	代数分式
45	term	项
46	like terms, similar terms	同类项
47	coefficient	系数
48	numerical coefficient	数字系数
49	literal coefficient	字母系数

3. 比率和比例

1	ratio	比率；比
2	rate	率；速率
3	per	每
4	vary	变化
5	proportion	比例
6	directly proportional to	成正比
7	inversely proportional to	成反比

4. 等式和不等式

1	equation	等式；方程
2	linear equation	线性方程；一次方程
3	quadratic equation	二次方程
4	system of equations	方程组
5	original equation	原方程
6	equivalent equation	同解方程

7	satisfy	满足（等式、不等式……）
8	inequality	不等式
9	triangle inequality	三角不等式
10	greater than	大于；多于
11	less than	小于；少于
12	greater than or equal to	大于或等于
13	less than or equal to	小于或等于

5. 函数

1	function	函数
2	domain	定义域
3	range	值域
4	interval	区间
5	define	定义（函数，集合等）
6	the definition of f	函数 f 的解析式
7	model	模型；表示
8	be modeled by the function f	……由函数 f 表示
9	graph	（函数或解析式的）图形
10	pass through	经过（某个点）
11	ordered pair	有序对
12	linear	线性的
13	collinear	共线的；同在一条直线上的
14	line	直线
15	parabola	抛物线
16	for which of the values of x	x 取哪一个值
17	attain	得到（……值）

6. 线和角

1	figure not drawn to scale	图形未按比例绘制
2	number line	数轴
3	segment/line segment	线段
4	labeled，indicated	标记的；标出的
5	tick marks	刻度线
6	be equally spaced	间隔相等
7	dashed line	虚线
8	ray	射线
9	length	长度
10	intersect	相交

11	bisect	平分
12	angle bisector	角平分线
13	extend	延长
14	perpendicular	垂直的；正交的
15	perpendicular line segment	垂直线段
16	parallel	平行线；平行的
17	transversal	截线
18	intercept	截距
19	vertex	顶点
20	midpoint	中点
21	endpoint	端点
22	angle	角
23	equiangular	等角的
24	measure	度数
25	number of degrees	度数
26	acute angle	锐角
27	right angle	直角
28	obtuse angle	钝角
29	straight angle	平角
30	round angle	周角
31	supplementary angle	补角
32	complementary angle	余角
33	adjacent angle	邻角
34	alternate angle	内错角
35	corresponding angle	同位角
36	vertical angle	对顶角
37	included angle	夹角
38	exterior angle	外角
39	interior angle	内角
40	central angle	圆心角

7. 三角形

1	triangle	三角形
2	oblique	斜三角形
3	scalene triangle	不等边三角形
4	isosceles triangle	等腰三角形
5	equilateral triangle	等边三角形
6	inscribed triangle	内接三角形
7	acute triangle	锐角三角形

8	obtuse triangle	钝角三角形
9	right triangle	直角三角形
10	side	边
11	hypotenuse	斜边
12	leg	直角边
13	Pythagorean theorem	勾股定理
14	altitude	高；高线
15	median of a triangle	三角形的中线
16	perimeter	周长
17	area	面积
18	base	底边；底面
19	similar（triangles）	相似（三角形）
20	radio of similitude	相似比
21	identical	恒等的；一样的
22	congruent	全等的

8. 四边形和多边形

1	quadrilateral	四边形；四边形的
2	parallelogram	平行四边形
3	rectangle	矩形；长方形
4	square	正方形
5	rhombus	菱形
6	trapezoid	梯形
7	polygon	多边形
8	regular polygon	正多边形
9	pentagon	五边形
10	hexagon	六边形
11	heptagon	七边形
12	octagon	八边形
13	nonagon	九边形
14	decagon	十边形
15	vertex	顶点
16	vertices	顶点（单数 vertex）
17	diagonal	对角线
18	opposite side	对边
19	opposite angle	对角
20	length	长
21	width	宽
22	nonagon	尺寸；维

9. 圆

1	circle	圆
2	semicircle	半圆
3	concentric circles	同心圆
4	circular	圆形的
5	center	圆心
6	radius	半径（复数 radii）
7	diameter	直径
8	chord	弦
9	circumference	圆周
10	arc	弧
11	segment of a circle	弧形
12	radian	弧度
13	curved path/curve	曲线
14	tangent	相切；相切的
15	inscribe	内接

10. 立体几何

1	solid	立体的；立体图形
2	rectangular	矩形的
3	rectangular solid/box	长方体
4	cube	立方体
5	cylinder	圆柱体
6	cylindrical	圆柱形的；圆柱体的
7	right circular cylinder	直圆柱体
8	sphere	球体
9	prism	棱柱体
10	cone	圆锥体
11	square pyramid	四角锥
12	triangular face	三角面
13	edge	棱
14	length	长
15	width	宽
16	height	高
17	face	面
18	plane	平面
19	cross section	横截面

20	volume	体积
21	surface area	表面积
22	cubic units	立方单位

11. 坐标几何

1	coordinate plane, xy-plane	坐标平面
2	axis	轴
3	x-axis	x 轴
4	y-axis	y 轴
5	coordinate	坐标
6	abscissa, x-coordinate	横坐标
7	ordinate, y-coordinate	纵坐标
8	quadrant	象限
9	origin	原点
10	point	点
11	be reflected across	以……为对称轴
12	reflection	对称（点，线）
13	respectively	分别地
14	lie	在（线或面）上
15	in the interior of	在……内部
16	distance	距离
17	x-intercept	与 x 轴的截距
18	y-intercept	与 y 轴的截距
19	horizontal	水平的
20	vertical	竖直的
21	slope	斜率
22	rise	（计算斜率时的）高
23	run	（计算斜率时的）长

12. 概率

1	probability	概率
2	event	事件
3	impossible	不可能的
4	certain	必然的
5	permutation	排列
6	combination	组合
7	arrange	排列
8	randomly, at random	任意地

| 9 | order | 顺序 |

13. 集合与数列

1	sequence	序列
2	term	项
3	the first term	第一项
4	the nth term	第 n 项
5	the preceding/previous term	前一项
6	arithmetic sequence	等差数列
7	geometric sequence	等比数列
8	element	元素
9	set	集合

14. 数据分析

1	figure	图形
2	grid	网格图；格子
3	pictogram/pictograph	象形图
4	bar graph	柱状图
5	circle graph, pie chart	饼状图
6	scatterplot	散点图
7	line of best fit	最适线
8	line graph	线形图
9	Venn diagram	维恩图
10	table	表格
11	shaded/darkened region	阴影部分
12	be consistent with	与……相符/一致
13	positive correlation	正相关
14	negative correlation	负相关

15. 单位

1	dollar	美元
2	cent	美分
3	penny	1美分硬币
4	nickel	5美分硬币
5	dime	1角硬币
6	dozen	打（12个）
7	score	廿（20）个
8	centigrade	摄氏

9	Celsius	摄氏
10	Fahrenheit	华氏
11	quart	夸脱（=2品脱）
12	pint	品脱
13	gallon	加仑（=4夸脱）
14	inch	英寸
15	foot	英尺（=12英寸）
16	yard	码（=3英尺）
17	meter	米
18	cubic meter	立方米
19	centimeter	厘米
20	micron	微米（百万分之一米）
21	hour	小时
22	minute	分钟
23	in dollars，inches，square feet	以美元、英寸、平方英尺为单位

16. 常见美制单位换算

长度	英寸：inch 英尺：feet/foot 码：yard	1 yard = 3 feet = 36 inches
重量	盎司：ounce 磅：pound 打兰：dram	1 pound = 16 ounce = 256 drams
体积	品脱：pint 加仑：gallon 夸脱：quart	1 gallon = 4 quarts = 8 pints
钱币	美元：dollar 美分：cent 10美分/一角：dime 5分镍币：nickel	1 dollar = 10 dimes = 20 nickels = 100 cent

附录3 数学计算器型号推荐购买列表

以下计算器均被 SAT 官方认可

序号	品牌	型号		
1	Casio/卡西欧	FX－6000 series FX－6500 series FX－7400 series FX－7800 series FX－8700 series FX－9750 series CFX－9850 series FX 1.0 series FX－CG－10(PRIZM)	FX－6200 series FX－7000 series FX－7500 series FX－8000 series FX－8800 series FX－9860 series CFX－9950 series Algebra FX 2.0 series FX－CG－20	FX－6300 series FX－7300 series FX－7700 series FX－8500 series FX－9700 series CFX－9800 series CFX－9970 series	
2	Hewlett-Packard/惠普	HP－9G HP－39 series HP－49 series	HP－28 series HP－40 series HP－50 series	HP－38G HP－48 series HP Prime	
3	Radio Shack/无线电器材公司	EC－4033 EC－4037	EC－4034		
4	Sharp/夏普	EL－5200 EL－9900 series	EL－9200 series	EL－9300 series	
5	Texas Instruments/美国德州仪器	TI－73　　TI－80 TI－83/TI－83 Plus TI－84 Plus TI－84 Plus C Silver TI－89 Titanium TI－Nspire/TI－Nspire CX TI－Nspire CAS/TI－Nspire CX CAS TI－Nspire CM－C/TI－Nspire CM－C CAS TI－Nspire CX－C CAS	TI－81　　TI－82 TI－83 Plus Silver TI－84 Plus CE TI－85　　TI－86	TI－84 Plus Silver TI－89	

致　谢

　　感谢在本书创作和出版过程中给予我们指导和帮助的所有人！

　　感谢父母的养育之恩和言传身教！感谢家人的默默支持！感谢吕蕾老师的专业指导和编辑的辛勤付出！感谢坚果教育南京和合肥同事们与老师们的协助、督促和鼓励！感谢坚果教育的李蓓老师为本书所做的资料整理和慷慨分享！感谢偏锋出国的金长麟老师和王悦老师以及坚果教育的吕广驰老师的积极参与和宣传、发行！

　　感谢 The Class of 2021（2017 年入学的一年级）的 Yale 大学的美高坚果代芮嘉；Dartmouth 学院的新加坡坚果徐珮瑶和南京坚果裘嘉西；Emory 大学的河南坚果李奉泰；UCLA 大学的江西坚果洪亦秋、合肥坚果周璞和西雅图坚果黄佳源；UC Berkeley 大学的南京坚果董安吉和昆明坚果柯辰昕；Brandeis 大学的北京坚果霍宇红；Carleton 文理学院的合肥坚果吴衡天和 Wheaton 文理学院的上海坚果王裴青以及上海坚果 Jason Zhu、Steven Lin、William Lu 和 Jimmy Yu，深圳坚果 Catherine Cao，兰州坚果 Harry Huang，南京坚果 Sophie Liu，洛杉矶坚果 Sharon Chen，圣地亚哥坚果 Tony Lin，波士顿坚果 Vivian Lee 和 Valentina Xu，这些同学在本书试用过程中提出了宝贵意见。

　　最后，更要感谢激发我们奋笔疾书反复研讨终成此书的万千学子们（坚果们）！愿这本书能助各位一臂之力，帮各位如你们的前辈那样早日圆高分梦、名校梦。"你是坚果，你也行！"

时　坚　言　中
2017 年 7 月于南京坚果教育

图书在版编目(CIP)数据

SAT-1数学轻松突破800分：思路与技巧的飞跃/时坚，言中编著. —北京：中国人民大学出版社，2017.7
ISBN 978-7-300-23954-5

Ⅰ.①S… Ⅱ.①时… ②言… Ⅲ.①数学-高等学校-入学考试-美国-教学参考资料 Ⅳ.①O1

中国版本图书馆CIP数据核字（2017）第013958号

SAT-1数学轻松突破800分：思路与技巧的飞跃
时 坚 言 中 编著
SAT-1 Shuxue Qingsong Tupo 800 Fen：Silu yu Jiqiao de Feiyue

出版发行	中国人民大学出版社	
社　　址	北京中关村大街31号	邮政编码　100080
电　　话	010-62511242（总编室）	010-62511770（质管部）
	010-82501766（邮购部）	010-62514148（门市部）
	010-62515195（发行公司）	010-62515275（盗版举报）
网　　址	http://www.crup.com.cn	
	http://www.1kao.com.cn（中国1考网）	
经　　销	新华书店	
印　　刷	北京鑫丰华彩印有限公司	
规　　格	185 mm×260 mm　16开本	版　次　2017年7月第1版
印　　张	13.75	印　次　2017年7月第1次印刷
字　　数	321 000	定　价　39.00元

版权所有　　侵权必究　　印装差错　　负责调换